Red Cell Manual

Red Cell Manual
Edition 7

Robert S. Hillman, MD
Chairman, Department of Medicine
Maine Medical Center
Portland, Maine
Professor of Medicine
University of Vermont College of Medicine
Burlington, Vermont

Clement A. Finch, MD
Professor Emeritus of Medicine
University of Washington School of Medicine
Seattle, Washington

F.A. DAVIS COMPANY • Philadelphia

F. A. Davis Company
1915 Arch Street
Philadelphia, PA 19103

Printed in the United States of America

Last digit indicates print number: 10 9 8 7 6 5 4 3 2 1

Medical Editor: Robert W. Reinhardt
Medical Developmental Editor: Bernice M. Wissler
Medical Production Editor: Jessica Howie Martin
Cover Designer: Robin Ricks, University of Washington
Cover Photograph: Dennis Kunkle, University of Hawaii

As new scientific information becomes available through basic and clinical research, recommended treatments and drug therapies undergo changes. The authors and publisher have done everything possible to make this book accurate, up to date, and in accord with accepted standards at the time of publication. The authors, editors, and publisher are not responsible for errors or omissions or for consequences from application of the book, and make no warranty, expressed or implied, in regard to the contents of the book. Any practice described in this book should be applied by the reader in accordance with professional standards of care used in regard to the unique circumstances that may apply in each situation. The reader is advised always to check product information (package inserts) for changes and new information regarding dose and contraindications before administering any drug. Caution is especially urged when using new or infrequently ordered drugs.

Library of Congress Cataloging-in-Publication Data

Hillman, Robert S., 1934–
Red cell manual / Robert S. Hillman, Clement A. Finch.—Ed. 7.
p. cm.
Includes bibliographical references and index.
ISBN 0-8036-0145-X
1. Erythrocyte disorders—Handbooks, manuals, etc. I. Finch, Clement A., 1915– . II. Title.
[DNLM: 1. Erythrocytes. 2. Anemia—diagnosis. 3. Anemia-therapy. 4. Erythropoiesis. 5. Hematologic Tests. 6. Polycythemia.
WH 150 H654r 1996]
RC647.E7H54 1996
616.1′51—dc20
DNLM/DLC
for Library of Congress 96-7350
CIP

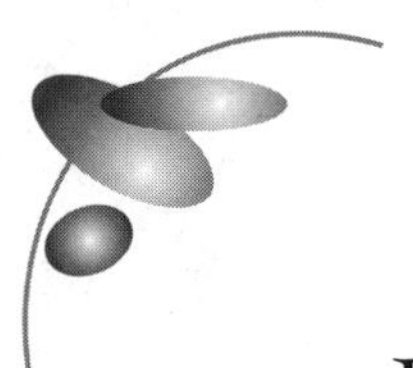

Preface

The ***Red Cell Manual*** is written for students, house officers, and physicians as a concise guide to the diagnosis and management of red cell disorders. It stresses a pathophysiological approach to diagnosis using commonly available laboratory tests. Basic concepts of erythron function are presented in the first part of the manual, followed by a discussion of the use of the laboratory in detecting and analyzing an anemia. The remainder of the manual is dedicated to a discussion of the functional approach to diagnosis and basic concepts of management. The manual is designed to provide an introduction to the area of red cell disease. For more extensive discussions of individual conditions, the student is referred to the literature citations and standard textbooks of hematology.

We wish to thank the many colleagues who have contributed over the years to the concepts presented in this manual. In the preparation of the seventh edition, special thanks must go to Arlene Shatz for her secretarial help, and to Dr. James R. McArthur and John Bolles of the University of Washington Educational Resources Center for the art work and the color photographs selected from the National Slide Bank of the American Society of Hematology.

Robert S. Hillman, MD
Clement A. Finch, MD

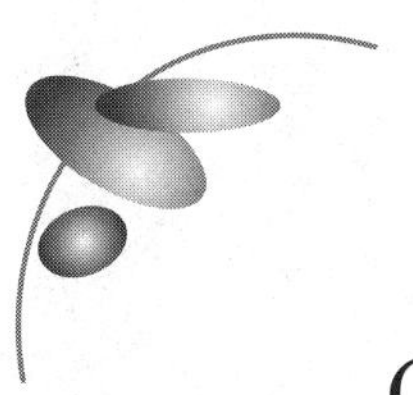

Contents

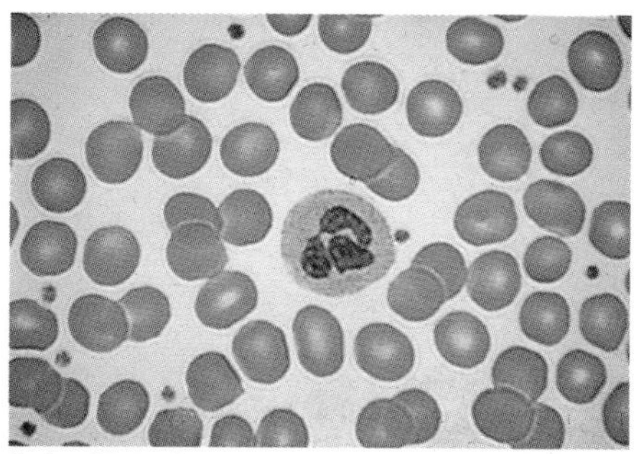

Color Plate 1. Normal peripheral blood smear (high-power magnification).

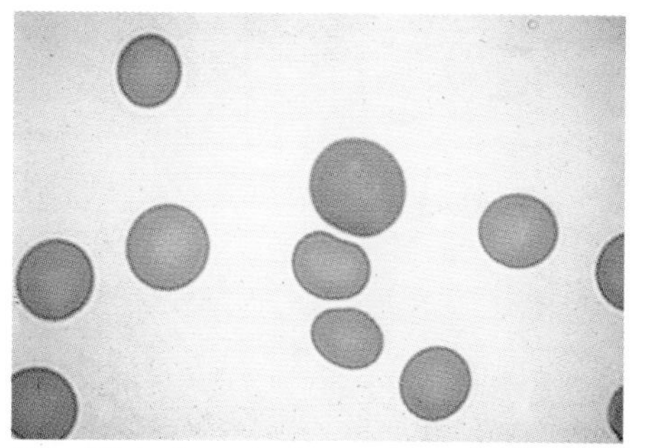

Color Plate 2. Polychromatic macrocyte (marrow reticulocyte).

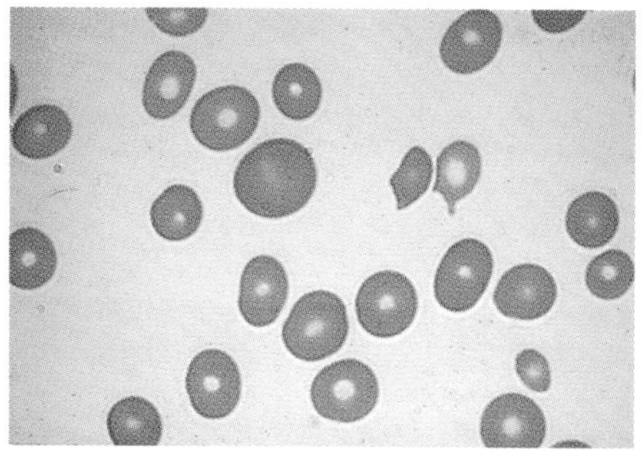

Color Plate 3. Anisocytosis.

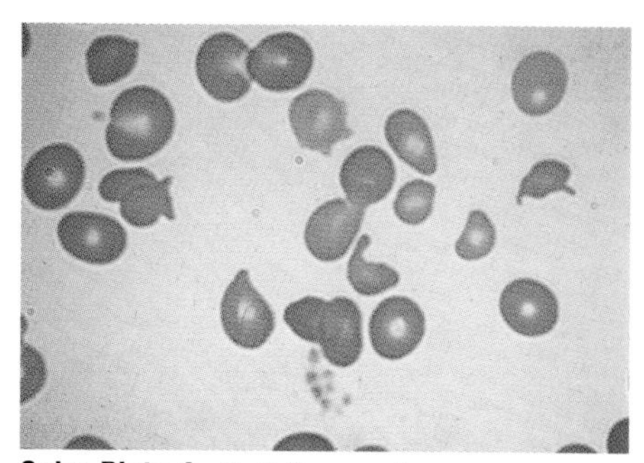

Color Plate 4. Poikilocytosis.

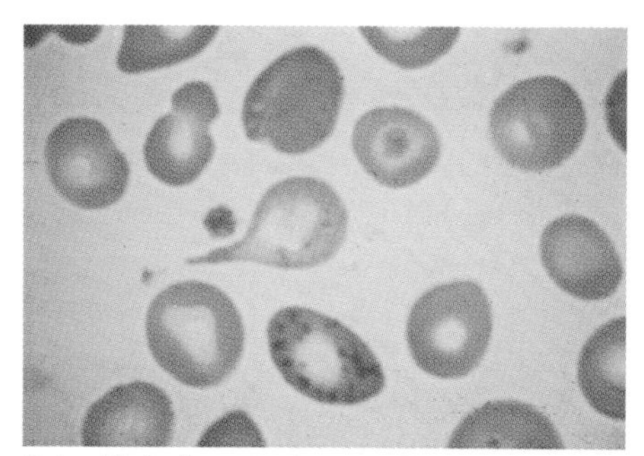

Color Plate 5. Stippling and a teardrop-shaped red blood cell.

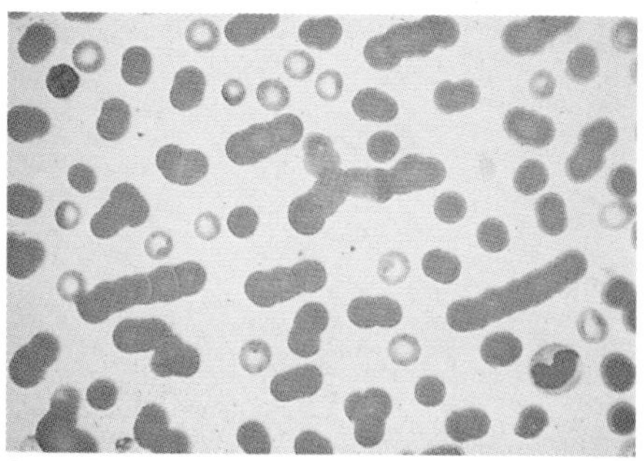

Color Plate 6. Rouleaux formation.

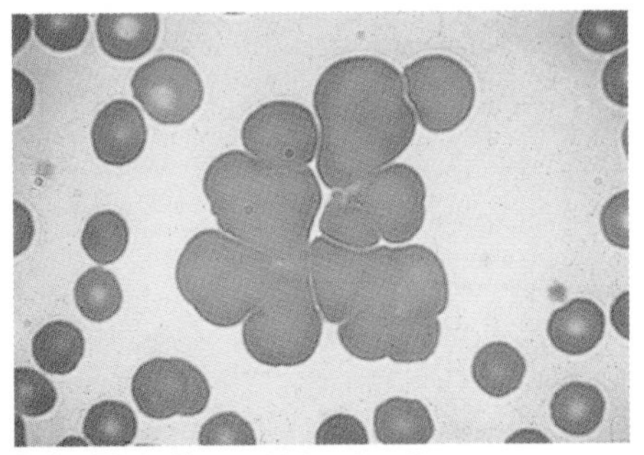

Color Plate 7. Red blood cell agglutination.

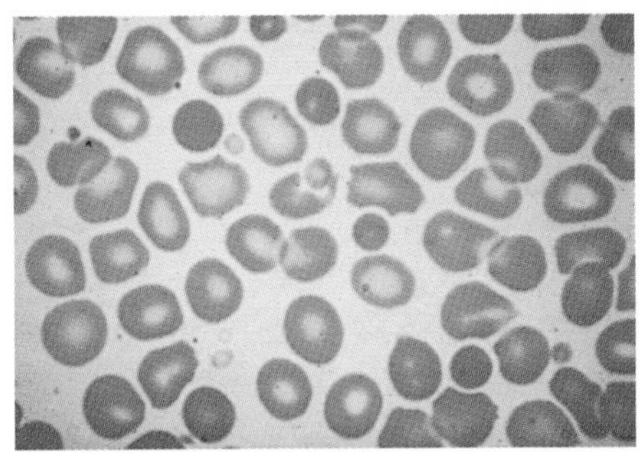

Color Plate 8. Spherocytosis.

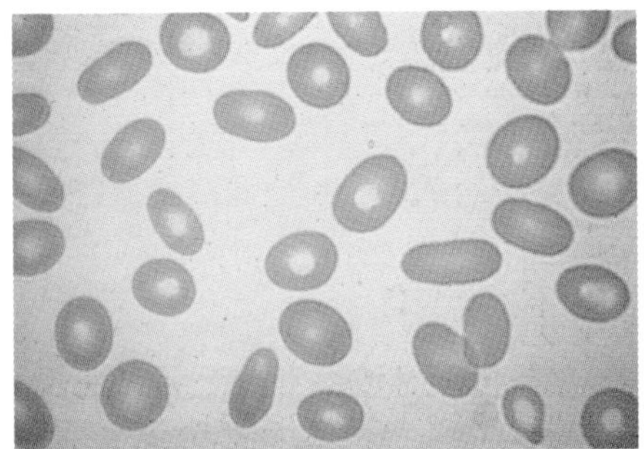

Color Plate 9. Elliptocytosis.

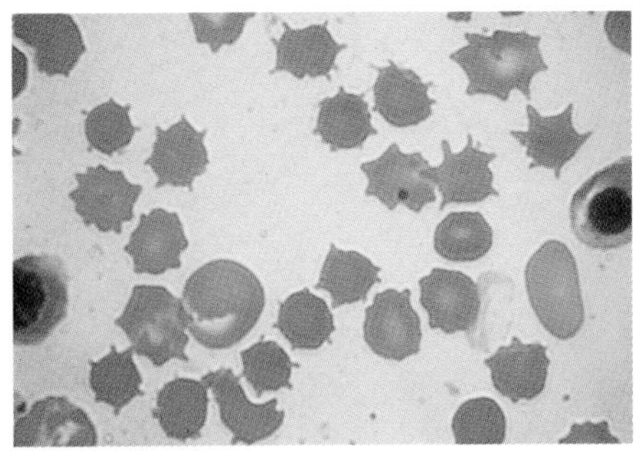

Color Plate 10. Acanthocytosis.

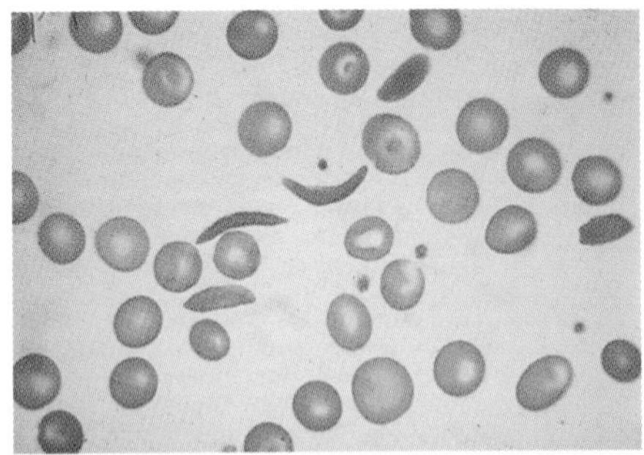

Color Plate 11. Sickle cells (Wright's stain).

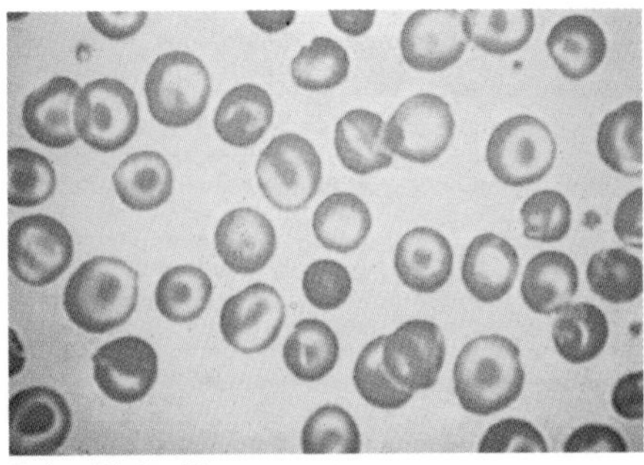

Color Plate 12. Targeting.

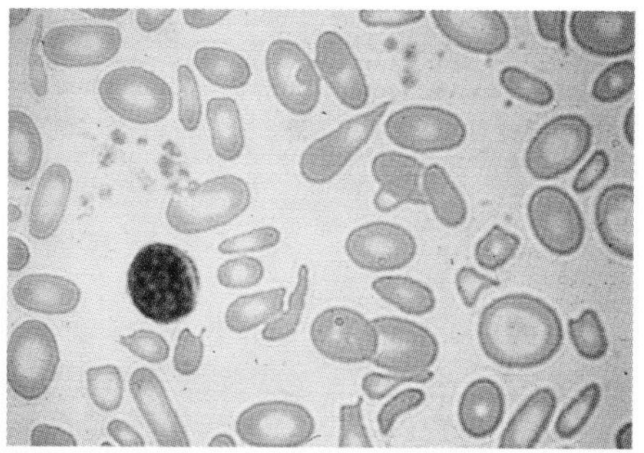

Color Plate 13. Severe iron deficiency with microcytosis, hypochromia, and marked anisocytosis and poikilocytosis.

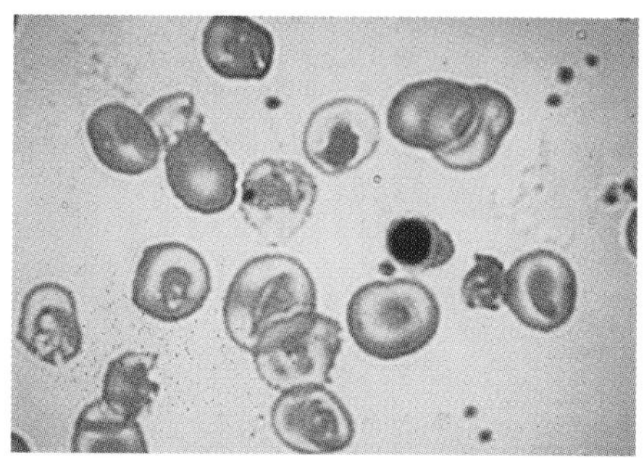

Color Plate 14. Thalassemia major with severe microcytosis, hypochromia, and targeting.

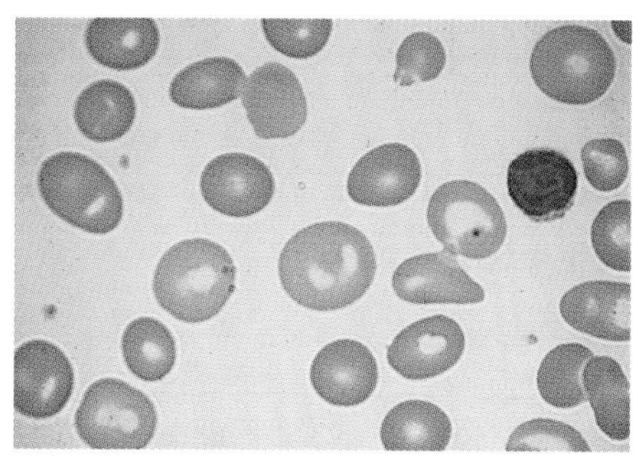

Color Plate 15. Macrocytosis.

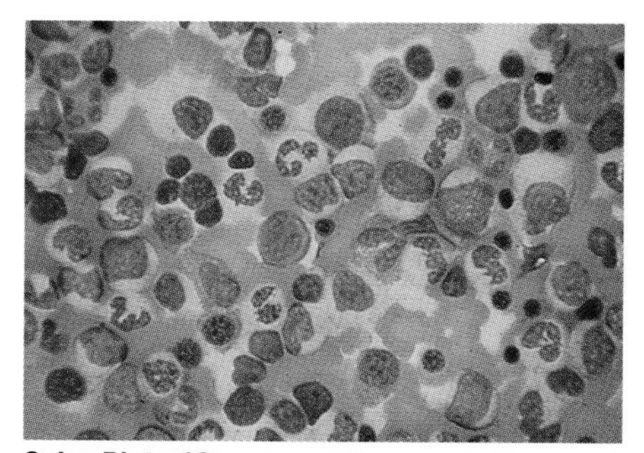

Color Plate 16. Normal bone marrow aspirate (E:G ratio of 1:3).

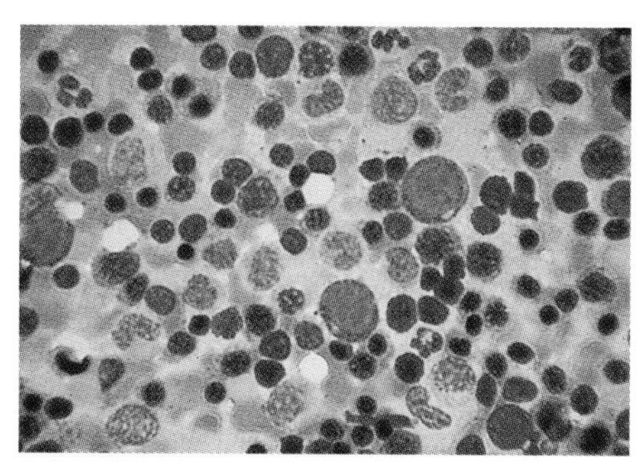

Color Plate 17. Erythroid hyperplasia (E:G ratio greater than 1:1).

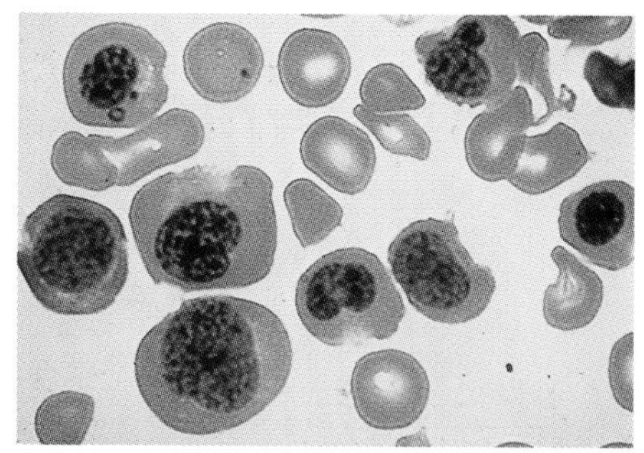

Color Plate 18. Megaloblastosis—high-power view of basophilic and orthochromatic megaloblasts with nuclear abnormalities.

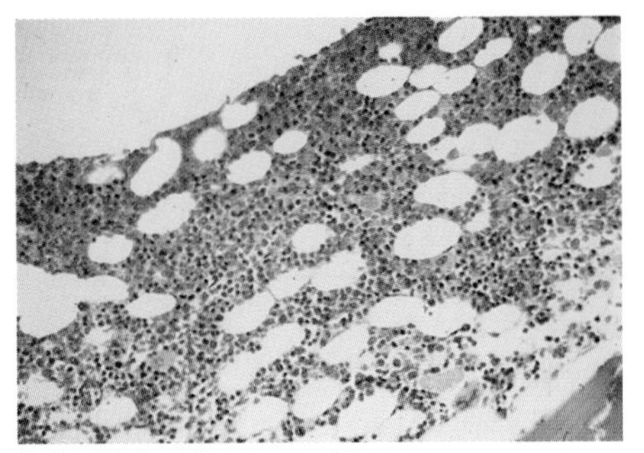

Color Plate 19. Normal bone marrow biopsy (lower-power magnification).

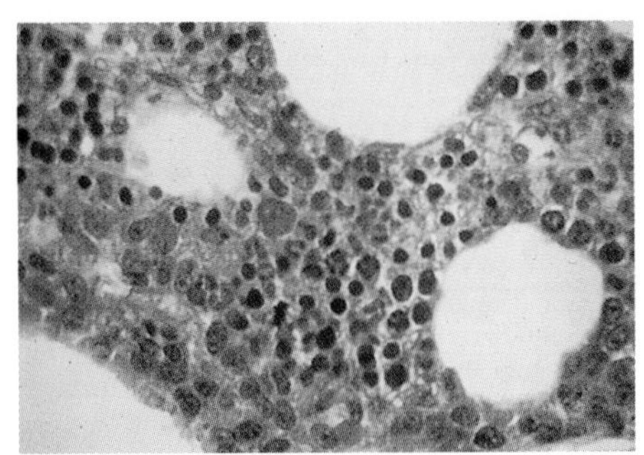

Color Plate 20. Normal bone marrow biopsy (high-power magnification).

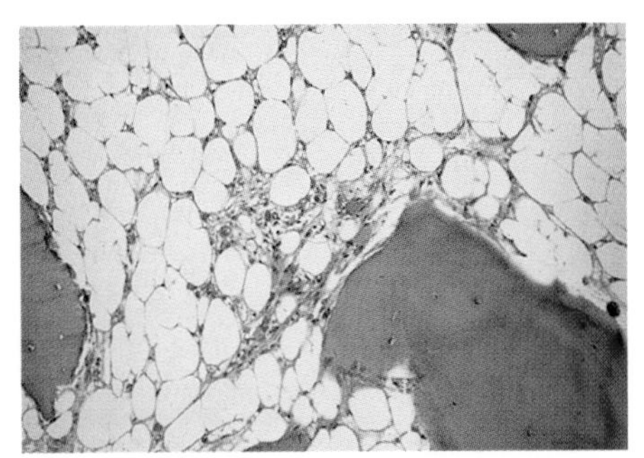

Color Plate 21. Aplastic anemia.

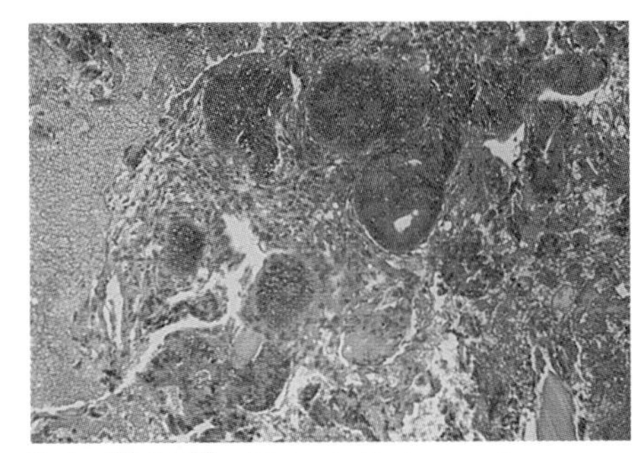

Color Plate 22. Metastatic tumor in a marrow biopsy.

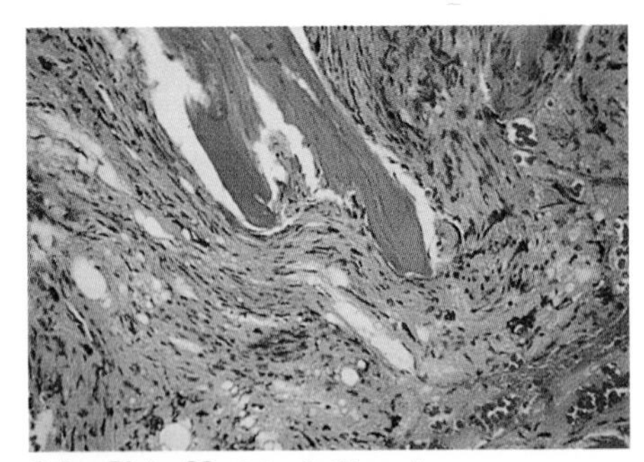

Color Plate 23. Myelofibrosis.

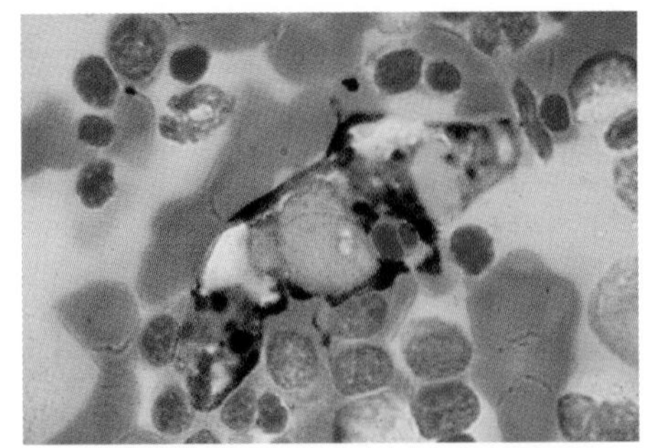

Color Plate 24. Reticuloendothelial cell stained with Prussian blue for iron stores.

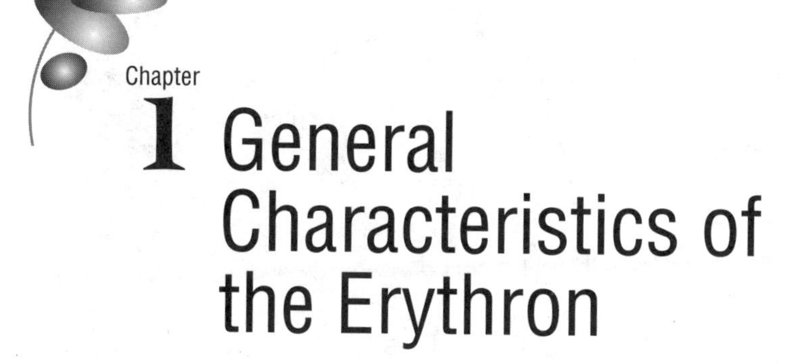

Chapter 1 General Characteristics of the Erythron

The human erythron is a highly specialized tissue that is responsible for oxygen transport. Marrow erythroid precursor cells, under humoral and cellular regulation, continually generate the required number of circulating mature erythrocytes. These cells are little more than containers for hemoglobin because they lack a nucleus and mitochondria and have barely enough metabolic machinery to defend themselves against the surrounding environment. However, the red cell is ideally suited for its oxygen-transport function. Assisted by various ligands, the hemoglobin within the cell releases oxygen at a suitable tension to support the energy-generating systems of body tissues. The first chapter of this manual reviews the general characteristics and function of the erythron as a background for discussing specific abnormalities of the erythroid marrow and circulating red cells.

Erythroid Marrow

Erythropoietic tissue originates in the yolk-sac mesenchyme, then moves to the liver and spleen during fetal life and, eventually, moves to its permanent home in the medullary cavity of the skeleton (Fig. 1). The adult distribution of the red cell marrow is limited to the axial skeleton and proximal ends of the long bones.[1] Although the supporting structure for the marrow persists in the more distal skeleton, hematopoietic tissue is replaced by fat cells in the adult. This process is reversible. Patients with stimulated erythropoiesis show both an increased number of erythroid cells in the

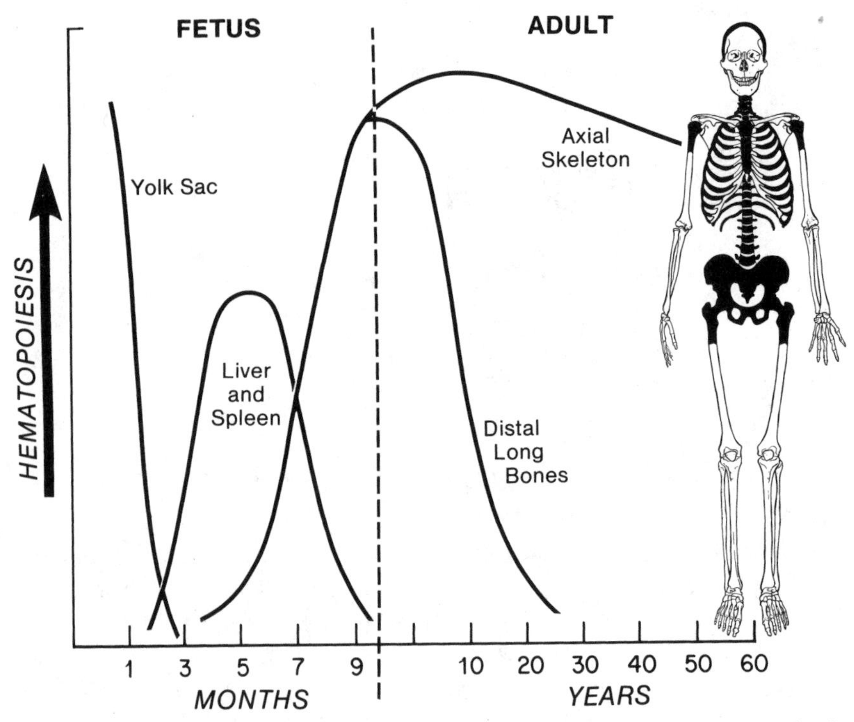

Figure 1. Location of active marrow growth in the fetus and adult. During fetal development, hematopoiesis is established in the yolk-sac mesenchyme, moves to the liver and spleen, and finally is limited to the bony skeleton. From infancy to adulthood, there is a progressive restriction of productive marrow to the axial skeleton and proximal ends of the long bones (*shaded areas*).

central marrow and a replacement of fat by active hematopoietic tissue in the peripheral skeleton.

The structure of the marrow provides a special environment for hematopoietic cell proliferation and maturation.[2] Hematopoietic cells are held in a matrix of reticular cells and fibers through which vascular sinusoids course, ending in a central venous sinus. The sinusoidal walls provide free access to plasma nutrients but retain developing cells until their rheologic properties permit penetration of the endothelial barrier. When a marrow aspirate is examined under the microscope, precursors for all hematopoietic cell lines appear to be mixed together in a jumbled mass. In-situ, however, cells tend to grow in clusters, with red cell precursors forming small erythropoietic islands. In the center of each is a macrophage.[3] This arrangement may be important for normal red cell maturation.

Genealogy of Erythroid Cells

All hematopoietic cell lines are derived from an original pool of pluripotent stem cells. These are semidormant cells with a lifelong capacity for self-renewal. Their morphologic characteristics are thought to resemble a small mononuclear, lymphocytelike cell. Although stem cells cannot be accurately identified on a marrow smear, their presence can be demonstrated experimentally by their growth characteristics in cell culture systems. When transplanted into a mouse whose marrow has been ablated by radiation, the pluripotent stem cells are capable of generating tumorlike clones of hematopoietic tissue in the marrow and spleen.

Studies using in vitro culture systems suggest that the pluripotent stem cell is the first in a sequence of steps of cell generation and maturation (Fig. 2).[4] This cell differentiates into either a lym-

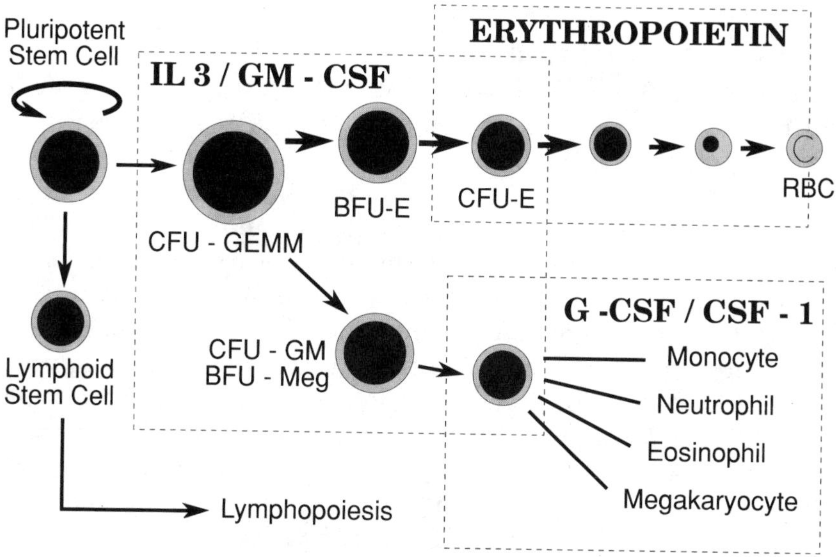

Figure 2. Sequence of marrow cell generation and maturation. A regenerating pool of pluripotent stem cells supports the multipotent stem cells required for erythropoiesis, myelopoiesis, and platelet formation and, along a separate pathway, for lymphopoiesis. As the stem cells proceed through a series of cell divisions, there is a progressive restriction in the cell lines they can support. The earliest identifiable cell in the red cell series, the burst-forming unit—erythroid (BFU–E), is controlled by growth conditioning factors (IL-3 and GM-CSF) derived from stromal cells, lymphocytes, and macrophages. The next step in differentiation is the formation of colony-forming unit—erythroid (CFU–E). This is under the control of the hormone erythropoietin.

phoid stem cell line to support lymphopoiesis or a multipotent stem cell (CFU–GEMM) capable of producing the other hematopoietic elements. In the case of erythropoiesis, the CFU–GEMM differentiates into a burst-forming unit—erythroid (BFU–E), the earliest identifiable cell committed to erythroid maturation. A single BFU–E is capable of producing in culture a visible colony containing more than 1000 erythroid cells. Proliferation of these early stem cells appears to be controlled by conditioning factors derived from lymphocytes and macrophages, including stem cell factor, interleukin-3 (IL-3), and granulocyte-macrophage colony-stimulating factor (GM-CSF). The next step in red cell differentiation is the formation of colony-forming units—erythroid (CFU–E). CFU–E produce erythroid colonies of up to 100 cells, in contrast to the larger coalescing colonies produced by the BFU–E. Subsequently, under the control of erythropoietin, these CFU–E undergo a programmed series of cell divisions and cell maturation leading to mature erythrocytes.

The Visible Erythroid Marrow

Although the BFU–E and CFU–E cannot be distinguished by light microscopic examination of stained smears of aspirated marrow, the progeny of the CFU–E are easily identified. Nucleated red cell precursors, called *normoblasts,* are distinguished from other primitive cells by their denser nuclear chromatin and lack of cytoplasmic granules, and, in later stages, by the appearance of hemoglobin within the cell cytoplasm (Fig. 3). Based on morphology with Wright's stain, the sequence of cell mitoses and maturation is divided into three phases: early, intermediate, and late cell maturation. Early maturation forms, the pronormoblasts and basophilic normoblasts, are large cells (300 to 800 fL) with slightly clumped nuclear chromatin that is heavier than that of white cells at a corresponding stage of maturation.[5] Nucleoli are generally not seen except at the earliest pronormoblast level, and even then are relatively indistinct. The cytoplasm is medium to dark blue and when viewed by light microscopy does not contain recognizable granules or organelles.

At the intermediate stage of maturation, normoblasts are smaller, with a more compact nucleus and some hemoglobin in the cytoplasm, imparting a bluish-green color. These cells are referred to as *polychromatic normoblasts.* In the late maturation stage, the nucleus continues to decrease in size and eventually becomes a dense,

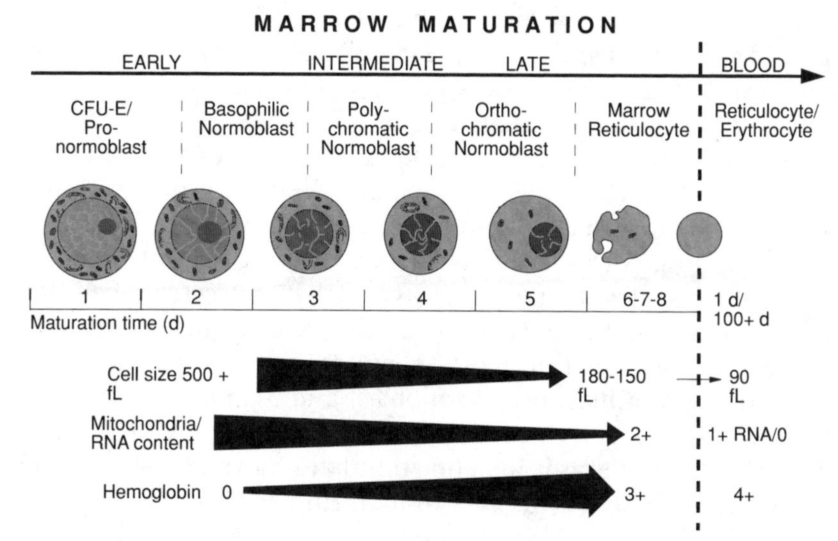

Figure 3. Erythroid precursor maturation. Maturation of the CFU–E to form adult erythrocytes can be divided into early, intermediate, and late stages of normoblast development. Using morphologic characteristics, this process can be further separated, as shown, into distinct populations of pronormoblasts: basophilic, polychromatic, and orthochromatic normoblasts and marrow reticulocytes. The maturation process takes 7 to 8 days and is associated with progressive formation of hemoglobin, a reduction in cell size, a loss of cellular mitochondria and RNA, and finally, a loss of the cell's nucleus. The final product of this process is the adult erythrocyte, a cell without a nucleus that survives in circulation for more than 100 days.

structureless mass. The cytoplasm is predominantly pink because of the increased hemoglobin content, and therefore the cells are referred to as *eosinophilic* or *orthochromatic normoblasts.*

The sequence is complete when the pyknotic nucleus is finally extruded from the cell, leaving a marrow reticulocyte, the immediate precursor of the circulating adult erythrocyte. By the time the nucleus is lost, the red cell contains about two-thirds of its eventual hemoglobin content.[6] The cell volume of 150 to 180 fL is still considerably larger than that of circulating mature erythrocyte. Moreover, the cytoplasm has a blue tinge caused by the presence of RNA, which along with residual mitochondria supports the synthesis of the remaining one-third of the cell's hemoglobin. Under normal circumstances, the marrow reticulocyte is held within the marrow while hemoglobinization continues, the content of RNA and mitochondria progressively decreases, and the cell shrinks. Finally, as its volume approaches that of a mature cell, it penetrates the sinusoidal wall and enters the circulation. However, it still contains a small amount of residual RNA for another 24 hours, allowing its identification in circulation as a newly formed erythrocyte (*blood reticulocyte*). These normal blood reticulocytes are usually not visible on Wright's stained smears, but they occasionally show a slight stippling caused by RNA precipitation.

Beginning with the differentiation of the CFU–E, the process of normoblast development involves 4 to 5 successive cell divisions, producing 16 to 32 adult erythrocytes from each pronormoblast (Fig. 4). The normoblast spends approximately 4 days as a proliferating and maturing normoblast and 3 additional days as a marrow reticulocyte. Under the stress of anemia or hypoxia, an increased output of erythropoietin stimulates more CFU–E to proliferate and increase the number of maturing erythroid precursors.[7]

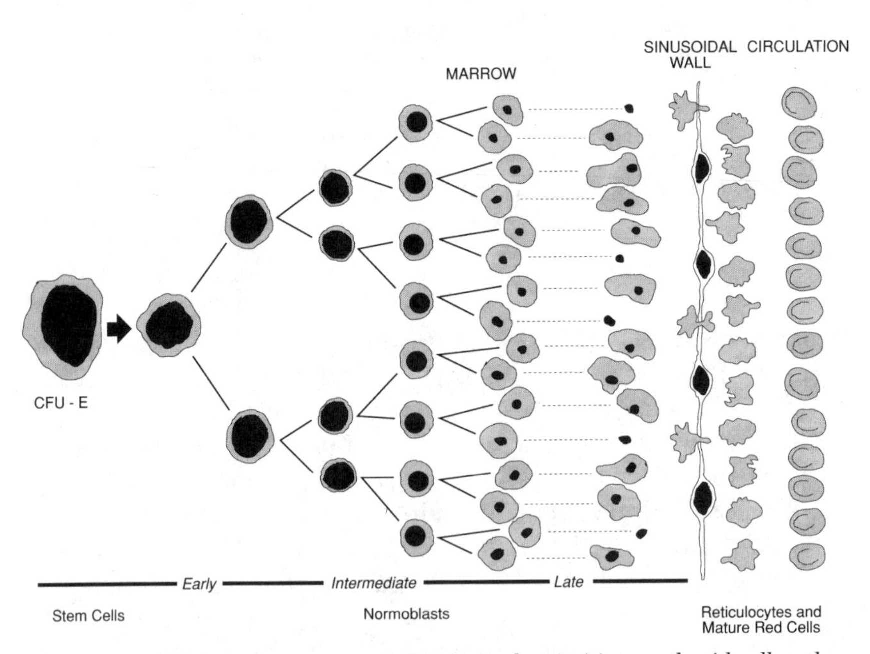

Figure 4. Proliferation sequence of CFU–E. Each primitive erythroid cell undergoes four mitoses, resulting in 14 to 16 progeny, which then mature and lose their nuclei to become marrow reticulocytes and, subsequently, circulating adult red blood cells. Cell death and removal by marrow reticuloendothelial cells during the process of maturation occurs to some extent even in normal marrows. In certain disease states, intramarrow destruction of maturing normoblasts can be greatly magnified, resulting in markedly ineffective erythropoiesis.

The total normoblast maturation time is shortened slightly, and more importantly, there is early delivery of marrow reticulocytes to circulation. This early delivery can shorten the response time to as little as 5 to 6 days. However, the extent of such changes depends on the functional integrity of the marrow, the severity and duration of the hypoxia or anemia, the adequacy of the erythropoietin response, and the availability of essential nutrients, especially iron.

Nutritional Factors in Erythropoiesis

Several nutritional factors are essential for normal erythropoiesis. Iron is uniquely essential because it is required for both proliferation and maturation of the developing red cell.[8] Hemoglobin synthesis is clearly dependent on the supply of iron, but more critical is cellular proliferation, which depends on sufficient iron supply. These effects are discussed in greater detail elsewhere in this chapter (see page 34).

Two vitamins, folic acid and vitamin B_{12}, are also required to support the integrated process of proliferation and maturation.[9] Both must be present in adequate amounts for normal methionine and thymidylate synthesis, key reactions required for normal DNA replication and sequential cell division. Methyltetrahydrofolate, the natural form of folate in most foods, acts as a methyl donor for the formation of methyl B_{12} (Fig. 5). The methyl group is then transferred to homocysteine to form methionine, an essential amino acid for protein metabolism. An inadequate supply of methyltetrahydrofolate disrupts both the formation of methyl B_{12} and subsequent steps in folate metabolism.

Levels of intracellular methyl B_{12} and a second active coenzyme, deoxyadenosyl B_{12}, can decrease when there is interference with the supply of vitamin B_{12}. The important impact of vitamin B_{12} deficiency is the failure of methyl groups to transfer from methyltetrahydrofolate to tetrahydrofolate. This failure traps dietary folate as methyltetrahydrofolate, and other folate congeners fall to levels insufficient to support normal DNA synthesis.

Folate and B_{12} deficiencies profoundly affect the maturation process of red cell precursors. Cells are enlarged, nuclei appear immature and are arrested in the S phase of mitosis, and cells at the normoblast stage undergo extensive destruction. This death of cells during development is called *ineffective erythropoiesis*.[10]

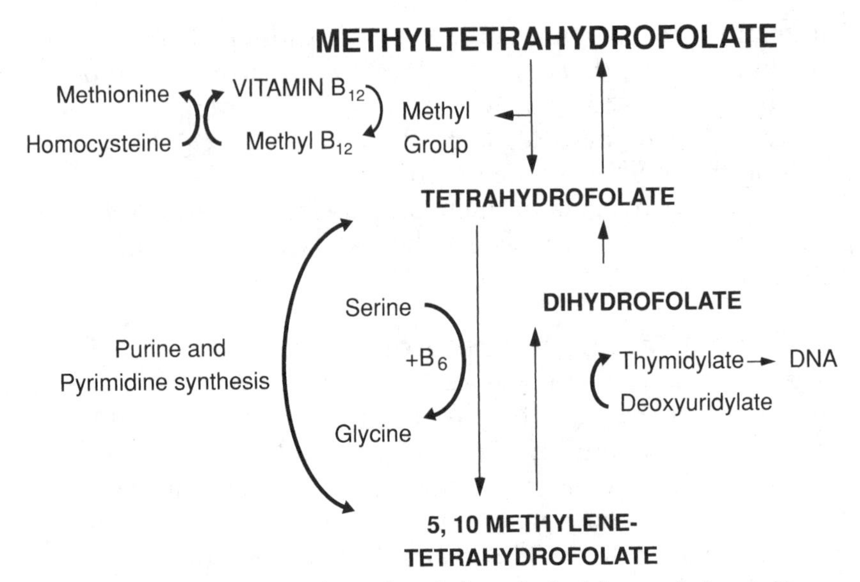

Figure 5. Metabolic relationships of methyltetrahydrofolate and vitamin B_{12}. Methyltetrahydrofolate is the methyl group donor for the formation of methyl B_{12}, an essential cofactor in the conversion of homocysteine to methionine. This generates tetrahydrofolate, which is the essential substrate for purine and pyrimidine synthesis and the conversion of deoxyuridylate to thymidylate for normal DNA synthesis. Tetrahydrofolate is also the natural substrate for the formation of folate polyglutamate in the liver.

Hemoglobin Synthesis

The immature erythroid cell is a dedicated factory for hemoglobin synthesis. This function requires the cell to have an adequate supply of iron as well as normal intracellular production of both porphyrin and the polypeptide chains of globin. Iron is conveyed through the plasma by the protein transferrin and is delivered to transferrin-iron receptors on the red cell membrane (Fig. 6).[11] Iron-transferrin-receptor complexes aggregate on the cell surface, causing the membrane to invaginate and form an intracytoplasmic vacuole. The iron is then released, and the receptor-transferrin complex returns to the cell membrane, whereupon the transferrin molecule reenters the plasma for further iron transport. The number of transferrin receptors on the cell surface determines the amount of iron taken up per cell, and the total number of cells determines the plasma iron clearance rate (turnover), if plasma iron is in adequate supply. After it is taken up by the cell, iron either enters the mitochondria for synthesis into heme, or, if present in excess, is stored as ferritin granules,

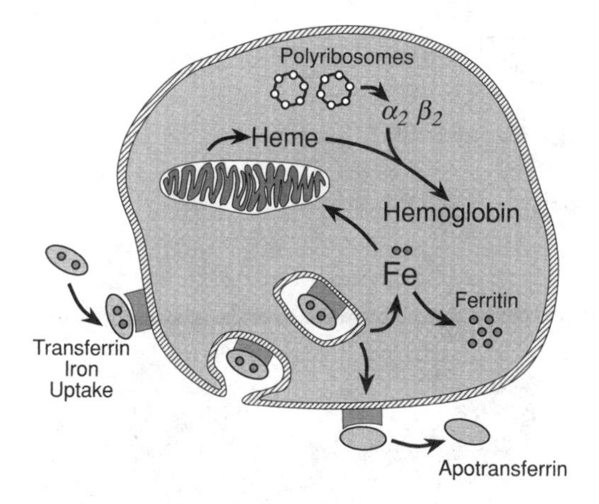

Figure 6. Intracellular pathways for iron uptake and incorporation into hemoglobin. The iron-transferrin complex is picked up by a membrane-associated receptor and brought into the cell by invagination and formation of an intracytoplasmic vacuole. The iron is then released and either stored as intracytoplasmic ferritin or used to synthesize heme, the precursor for hemoglobin. The transferrin receptor complex is returned to the cell membrane, where the apotransferrin is expelled back into circulation.

which are semicrystalline iron-protein aggregates. Cells containing ferritin-iron granules *(sideroblasts)* can be visualized on marrow aspirates or biopsies using a specific iron stain.

Porphyrin synthesis begins in the mitochondria with the formation of δ-aminolevulinic acid (ALA) from glycine and succinyl-coenzyme (succinyl-CoA) and then shifts to the cytoplasm for the production of porphobilinogen (Fig. 7).[12] Four molecules of porphobilinogen are assembled in a ring structure to form in sequence uroporphyrinogen and coproporphyrinogen. The final step occurs within the mitochondria; protoporphyrin forms and incorporates with iron to form heme. Normally, a small amount of excess porphyrin complexed to zinc remains after heme synthesis is complete. This excess is referred to as *free erythrocyte protoporphyrin* (FEP). The amount of FEP detectable in circulating red blood cells increases when iron supply is inadequate.

Globin, the protein portion of the hemoglobin molecule, is a tetramer of polypeptide chains encoded by two clusters of closely linked genes on chromosomes 16 (α-type genes) and 11 (non-α genes) (Fig. 8). The type of chain produced in fetal life is altered soon after birth as a result of the sequential suppression and activation of individual genes. At birth, red cells contain mainly fetal hemoglobin (Hb F), which is made up of two α chains and two γ chains. Within a few months after birth, Hb F largely disappears and is replaced by adult hemoglobin (Hb A). This tetramer is composed of two α chains,

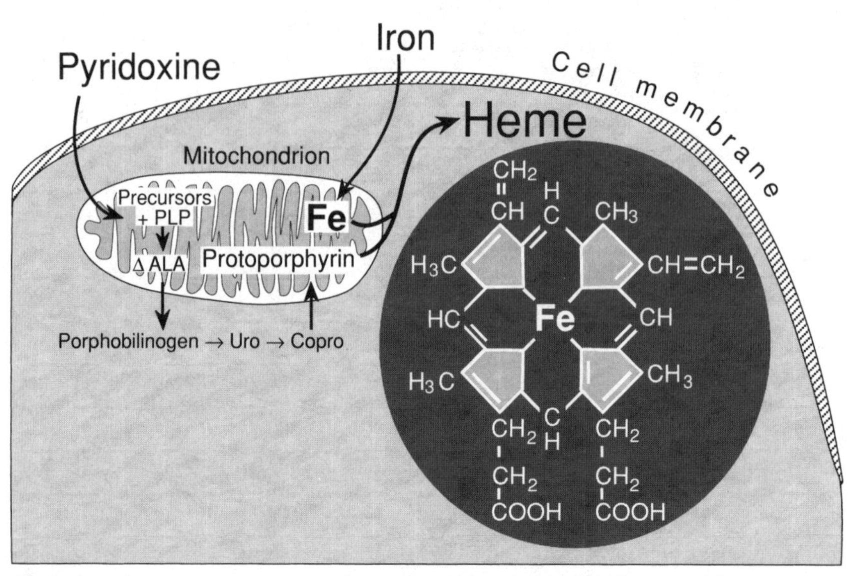

Figure 7. Heme formation. The mitochondrion is responsible for the synthesis of protoporphyrin, a stepwise process beginning with the formation of δ-aminolevulinic acid from glycine and succinyl-CoA with pyridoxal 5-phosphate (PLP) as an essential cofactor. The sequence of porphobilinogen, uroporphyrin, and coporphyrin formation then occurs in the cytoplasm, followed by an intramitochondrial assembly of protoporphyrin and iron to form heme. The structure of the final product, the heme molecule, is shown. It consists of four porphyrin moieties assembled in a ring structure around a central iron molecule.

each containing 141 amino acids, encoded by the closely linked α genes on chromosome 16, and two β chains with 146 amino acids, encoded by the single β-chain gene on chromosome 11. Hemoglobin A makes up 96% to 97% of the hemoglobin in adult erythrocytes. Two minor hemoglobins, hemoglobin F (less than 1%) and hemoglobin A_2 (approximately 2.5%), are also present. The pattern of globin synthesis is important because changes in chain concentration provide clues to disorders of globin synthesis. Globin gene mutations can change the structure, function, and production of hemoglobin to produce a number of different hemoglobinopathies.

The Mature Red Cell

The mature red cell is a biconcave disk with a mean diameter of 8 μm, a thickness of 2 μm, and a volume of 90 fL (Fig. 9).[13] The cell is devoid of a nucleus and mitochondria and has lost the ability to synthesize protein. Its limited metabolism is barely enough to sustain it

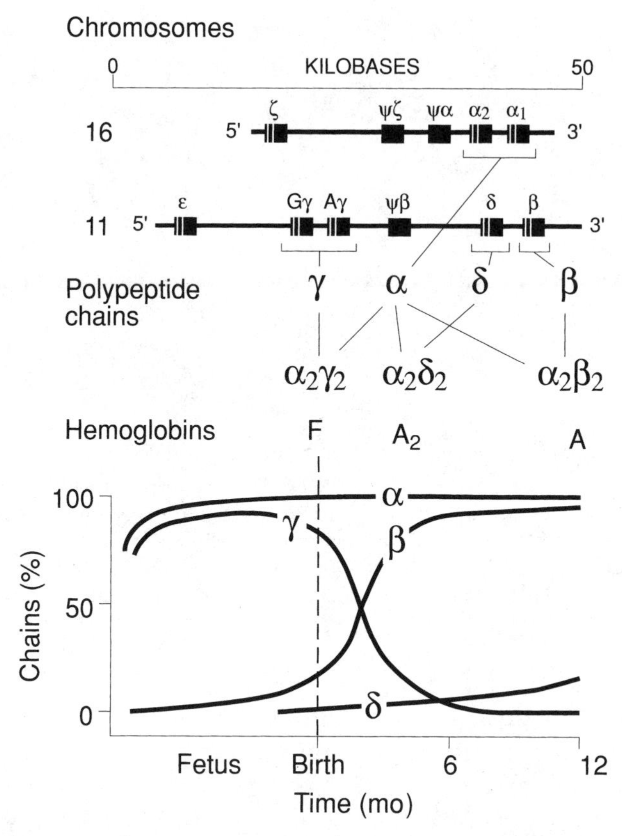

Figure 8. Changes in hemoglobin with development. Sequential suppression and activation of individual globin genes in the immediate postnatal period results in a switch from fetal hemoglobin (Hb F: 2-α and 2-γ chains) to adult hemoglobin (Hb A: 2-α and 2-β chains). A small amount of Hb A_2 (2-α and 2-δ chains) is also present in the adult.

during its 4-month life span in circulation. However, the enucleated red cell is admirably designed to survive many trips through the microvasculature. It is resilient and capable of extreme changes in shape, and its hemoglobin supports efficient oxygen transport and delivery. The proper function and longevity of the erythrocyte depend on the relationships between the cell membrane and metabolism.

Membrane

The limiting barrier between the red cell cytoplasm and the environment is a lipid sheath, two molecules thick, consisting of phospholipid molecules packed tightly together with their polar ends

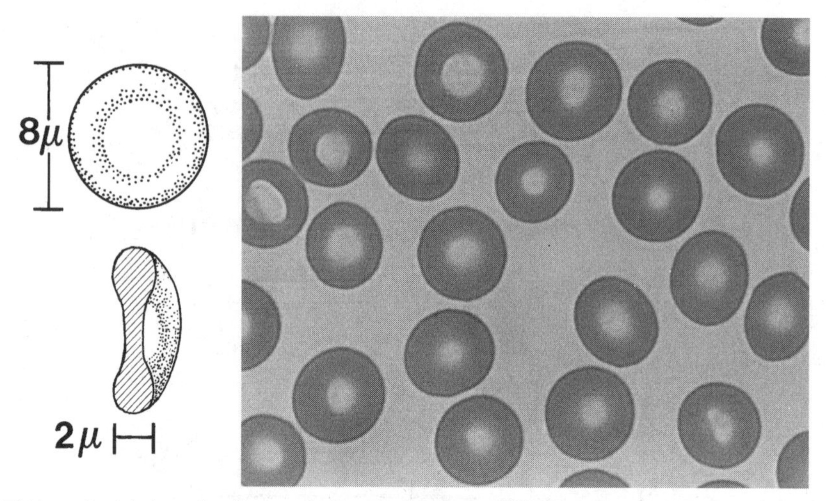

Figure 9. Adult red cell morphology. The normal adult red cell has a characteristic biconcave disk shape. By light microscopy (using a dry blood smear stained with Wright's stain), adult red cells appear as cells without nuclei, which are generally round with a lighter staining central area, compatible with the biconcave shape.

facing the aqueous phase on either side of the membrane (Fig. 10).[14] The external surface is relatively rich in phosphatidylcholine, sphingomyelin, and glycolipid; the inner portion of the membrane contains mostly phosphatidylserine, phosphatidylethanolamine, and phosphatidylinositol. Cholesterol is present in a 1:1 molar ratio with the phospholipids and is in rapid exchange with the unesterified cholesterol of the plasma. The cholesterol content of the membrane depends on the concentrations of free plasma cholesterol and bile acids and the activity of the esterifying enzyme, lecithin-cholesterol acyltransferase (LCAT).

Changes in red cell shape can result from changes in plasma lipids, and survival is variably affected. Patients with hepatocellular disease or biliary obstruction show impaired LCAT activity, resulting in cholesterol overloading of the red cell membrane. Such red cells have excess membrane, which makes them appear as target or spiculated cells on the blood smear. Spiculated cells (*acanthocytes*) are also observed in patients who lack a β lipoprotein in their plasma. As a consequence, they show major distortions in the ratios of membrane lecithins and sphingomyelins with an associated shortened red cell survival. Other abnormalities of membrane phospholipids can reduce cell survival without a specific morphologic change.

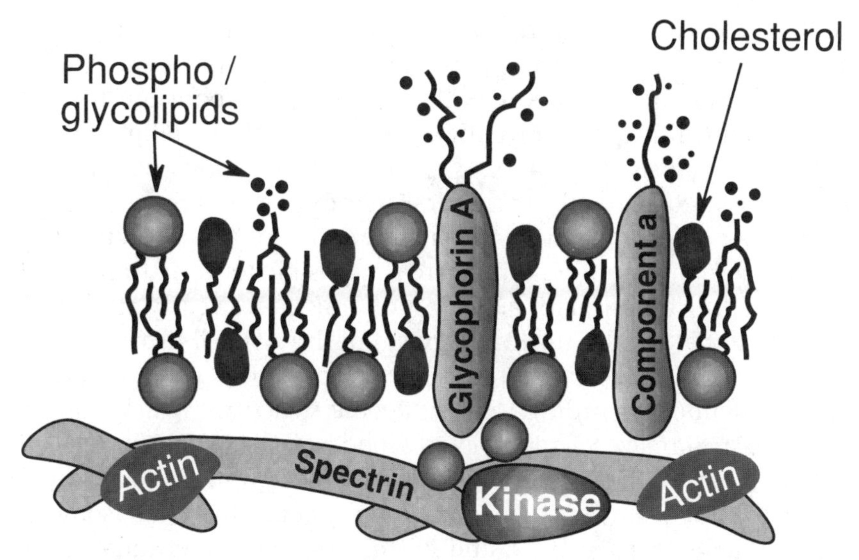

Figure 10. Structure of the red cell membrane. The adult red cell membrane consists of lipid sheath, two molecules thick, affixed to a reticular protein network made up of two major components, spectrin and actin. The external surface is relatively rich in phosphatidylcholine, sphingomyelin, and glycolipid, whereas the inner lipid layer contains mostly phosphatidylserine, phosphatidylethanolamine, and phosphatidylinositol. These phosphoglycolipids are present in an equimolar ratio with cholesterol. In addition, glycophorin A and component a (band 3), two integral proteins, penetrate the bimolecular lipid sheath and are in contact with the aqueous phase on either side. The membrane proteins provide a negative charge to the surface of the cell and help in the diffusion of glucose and anions across the membrane.

As illustrated in Figure 10, about half of the membrane mass is made up of two classes of protein.[15] The *integral proteins,* principally glycophorin A and component a (band 3), penetrate the bimolecular lipid sheath and are in contact with the aqueous phase on either side. About 60% of the glycophorin A molecule is made up of carbohydrate as oligosaccharide chains. These are attached to the outside end of the molecule and serve to provide a negative charge to the surface of the red cell, an important characteristic for preventing cell agglutination. Component a (band 3) may be involved in the diffusion of glucose and anions across the membrane.

A second class of protein, the *peripheral proteins,* form a reticular network on the inner surface of the membrane. Two components, spectrin and actin, are in greatest abundance. They form a lattice, which is attached to the cytoplasmic ends of the integral proteins, fixing their position and creating a cell skeleton. The red cell's biconcave shape and many of its mechanical properties may well be determined by the peripheral membrane proteins. An abnormality in these proteins is thought to be responsible for the shape deformities seen in patients with hereditary elliptocytosis and spherocytosis. A number of specific receptors and enzymes are also associated with the peripheral membrane proteins. Some of these may be important in the maintenance of the structural integrity of the membrane; others are responsible for procurement of essential nutrients and for active cation transport. Nearly all appear to be essential for the normal survival of the red cell in vivo.

The character of the external surface of the red cell membrane may be defined according to its antigenic structure, which has resulted in a complex terminology of blood types.[16] Over 300 red cell antigens have been identified; many of these antigens constitute about 15 genetically distinct blood group systems. In at least some of these systems (e.g., A, B, O, and P), the antigenic determinants are composed largely of the oligosaccharide prosthetic groups of the integral membrane proteins and complex glycosphingolipids. Nearly all antigens are intrinsic components of the membrane, appearing during the cell's early development. The Lewis system is an exception because its antigens are on glycolipids present in tissue fluids and are only secondarily absorbed onto red cells. Other passenger "antigens" that can appear on the red cell surface under pathologic conditions include bacterial polysaccharides and certain drugs such as penicillin.

As with the other components of the red cell membrane, changes in antigenic structure may or may not be associated with cell-shape abnormalities. Cells that lack any of the Rh antigens, *Rh null cells,* appear as stomatocytes on the peripheral blood smear and have a shortened survival rate.[17] Furthermore, in the presence of antibodies against specific red cell surface antigens, there can be increased cell destruction with or without morphologic changes. The coating of a surface antigen with antibody not only interferes with membrane function and integrity, but most importantly permits rapid phagocytosis of the altered red cells by reticuloendothelial cells equipped with receptors for the immunoglobulin G (IgG) Fc fragment and C3b component of complement.

Red Cell Metabolism

Because the mature red cell lacks a nucleus, its metabolism is unique. In the absence of mitochondria, the cell has little ability to metabolize fatty acids and amino acids. Energy is generated almost exclusively through the breakdown of glucose (Fig. 11). This meta-

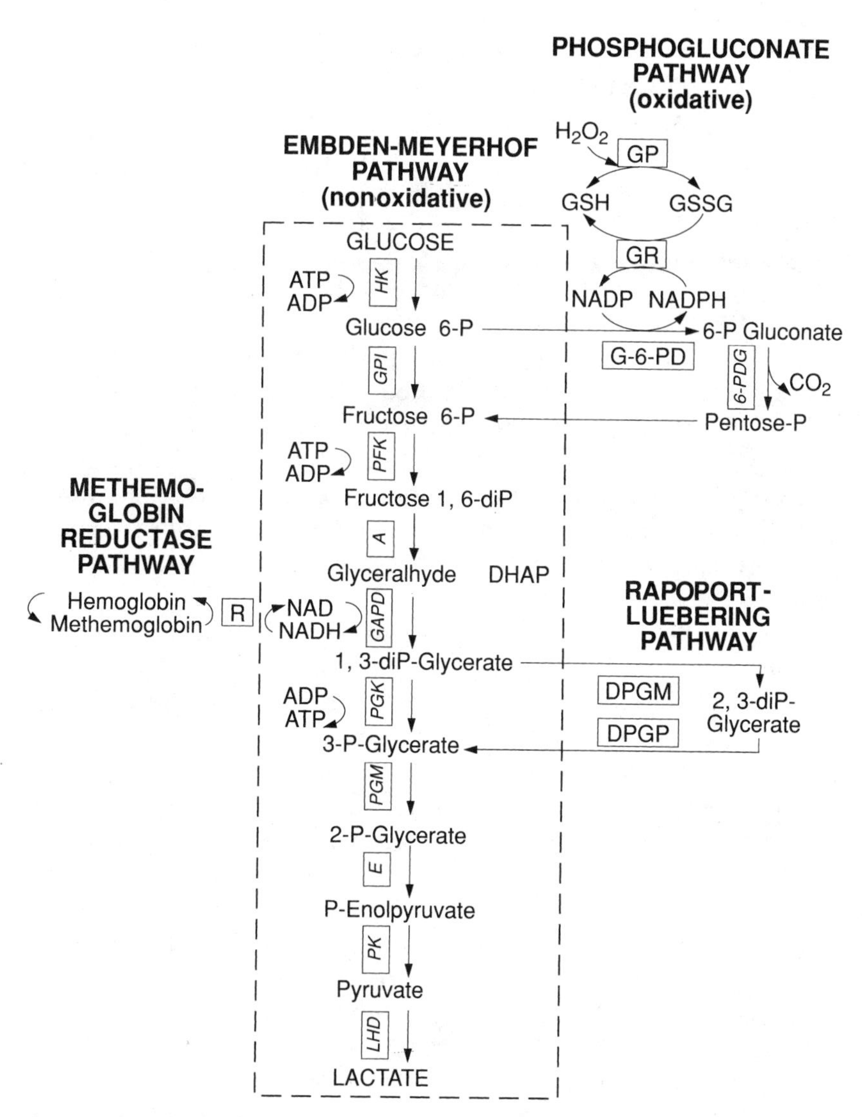

Figure 11. Red cell metabolic pathways. The enucleated red cell depends almost exclusively on the breakdown of glucose for energy requirements. The Embden-Meyerhof pathway (nonoxidative or anaerobic pathway) is responsible for most of the glucose use and generation of adenosine 5′-triphosphate. In addition, this pathway plays an essential role in maintaining pyridine nucleotides in a reduced state to support methemoglobin reduction (the methemoglobin-reductase pathway) and 2,3-diphosphoglycerate (2,3-DPG) synthesis (the Rapoport-Luebering pathway). The phosphogluconate pathway couples oxidative metabolism with pyridine nucleotide and glutathione reduction and protects red cells from environmental oxidants.

bolic activity can be conveniently subdivided into the major anaerobic (Embden-Meyerhof) pathway and three ancillary pathways. All of these pathways are closely related and must function adequately if the erythrocyte is to transport oxygen normally and survive in circulation.

The Embden-Meyerhof pathway is responsible for about 90% of the cell's glucose utilization. In the conversion of glucose to lactate, 2 mol of adenosine 5′-triphosphate (ATP) are consumed during the hexose portion of the pathway, but 3 to 4 mol are generated at the triose level. It is this net gain in ATP that provides the high-energy phosphate necessary for maintaining cell shape and flexibility, preserving membrane lipids, and energizing the metabolic pumps that control sodium, potassium, and calcium flux. When ATP is deficient because of an inherited or acquired defect in glycolysis, cell survival is dramatically reduced, and a hemolytic anemia results. The loss of intracellular ATP in red cells stored for several weeks in the blood bank is responsible for the loss of their viability after transfusion.

The methemoglobin-reductase pathway depends on the Embden-Meyerhof pathway for the reduced pyridine nucleotides, which keep hemoglobin in a reduced state. Methemoglobin, resulting from the conversion of heme iron from the ferrous to the ferric form, cannot combine reversibly with oxygen.[18] Faced with the constant oxidant stress of the environment, an unchecked formation of methemoglobin would greatly reduce the oxygen-carrying capacity of the red cell. The methemoglobin-reductase pathway counteracts the oxidant state by reducing the hemoglobin iron to its ferrous form. The pathway functions through the enzyme methemoglobin reductase and the reducing capacity of pyridine nucleotide (NAD). Patients who are homozygous for an abnormal methemoglobin-reductase gene show a complete failure of conversion of methemoglobin to hemoglobin and accumulate between 20% and 40% methemoglobin in circulating red cells. In the heterozygous state, the partial enzyme deficiency is sufficient to maintain reduced hemoglobin under normal conditions, but challenge by an oxidant drug produces methemoglobinemia caused by a lower rate of conversion of methemoglobin to hemoglobin.[19]

The Rapoport-Luebering shunt, necessary for the production of 2,3-diphosphoglycerate (2,3-DPG), is also an offshoot of the anaerobic glycolytic pathway. The amount of 2,3-DPG produced depends primarily on the rate of glycolysis as determined by the enzyme phosphofructokinase.[20] This enzyme is pH sensitive, increasing the rate of glycolysis as the pH increases and decreasing the rate as the pH falls. The DPG response is also dependent on an adequate supply of inorganic phosphate. Phosphate depletion in certain nutritional plasma states and following treatment of diabetic acidosis may greatly retard the production of DPG. The significance of 2,3-DPG lies in its ability to modulate oxygen release according to tissue needs. This response is activated by a change in the proportion of oxygen extracted by tissues; whenever venous blood contains an increased proportion of deoxygenated hemoglobin, glycolysis is stimulated to produce more DPG. This decreases the affinity of hemoglobin for its oxygen and allows more oxygen to be released at any given tissue-oxygen tension.

The other pathway is the hexose monophosphate shunt (phosphogluconate pathway). This ancillary energy system couples oxidative metabolism with pyridine nucleotide (NADP) and glutathione reduction. Its activity increases with increased oxidation of glutathione. When the pathway is functionally deficient or environmental oxidants exceed its reducing capacity, globin denaturation occurs, and hemoglobin precipitates to form Heinz-Ehrlich bodies along the inner surface of the erythrocyte membrane. Such precipitates are thought to produce membrane damage. This type of oxidative destruction of red cells usually occurs in patients with x-linked glucose-6-phosphate-dehydrogenase deficiency, a principal enzyme in the phosphogluconate pathway (see page 115).

Hemoglobin

Hemoglobin is the main intracellular protein of the red cell, constituting approximately 33% of its contents (Fig. 12). It has a molecular weight of 68,000 d and is assembled from four polypeptide chains, each containing a heme group in a hydrophobic pocket. The hemoglobin molecule undergoes conformational changes as it interacts with various ligands, including oxygen, hydrogen, carbon dioxide and 2,3-DPG.[21] In the deoxyhemoglobin form, hydrogen ions establish salt bridges between the individual chains, while carbon dioxide and 2,3-DPG bind the two β chains to restrain the hemoglobin in a low-affinity form (Fig. 13). With oxygen uptake, subtle shifts in tertiary structure occur, causing rupture of salt bonds and apposition of the β chains to expel the 2,3-DPG and carbon dioxide from the crevice between them. This respiratory motion increases the affinity of hemoglobin for oxygen and is responsible for the sigmoid form of

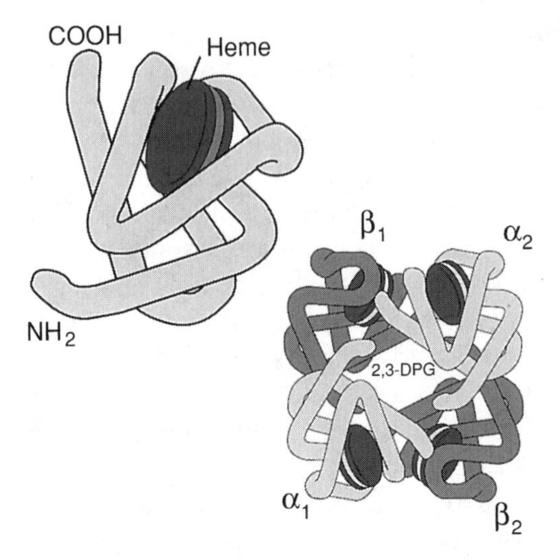

Figure 12. Hemoglobin structure. Hemoglobin molecules are composed of four subunits: two α chains (*light shading*) and two β chains (*dark shading*). Each of the globin chains provides a pocket for a heme molecule and, therefore, has the capacity to bind up to four molecules of oxygen. 2,3-DPG binds to the two β chains to stabilize the molecule when it is in the deoxygenated state.

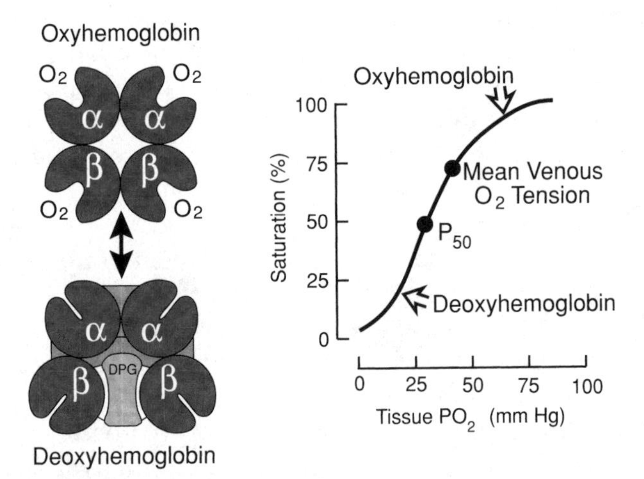

Figure 13. Hemoglobin function, as expressed by the hemoglobin-oxygen dissociation curve. The hemoglobin molecule changes its conformation as it loads and unloads oxygen and carbon dioxide. In the deoxygenated state, the molecule opens to accept a single molecule of 2,3-DPG, and hydrogen ions establish salt bridges between individual chains. With oxygen uptake, the salt bonds are ruptured, the 2,3-DPG and carbon dioxide are expelled, and the heme groups open to receive four molecules of oxygen. This unique respiratory motion of the hemoglobin molecule is responsible for the sigmoid shape of the hemoglobin-oxygen dissociation curve.

the oxygen dissociation curve. Clinically, the hemoglobin affinity determines the proportion of oxygen that will be released or loaded at any given oxygen tension (torr). Higher or lower affinities are seen in patients with certain abnormalities of hemoglobin structure.

Qualitative abnormalities of the hemoglobin molecule are mostly caused by genetically induced substitutions of single amino acids in its α or β polypeptide chains. Depending on the position of the substitution, a variety of effects may be produced. Some substitutions alter the affinity of hemoglobin for oxygen by interfering with the normal intramolecular movement from a low- to a high-affinity state. Other substitutions involve a change in the valency of heme iron from ferrous to ferric. Generally, these substitutions represent a defect in a polypeptide chain in the vicinity of one of the heme groups. A number of substitutions result in instability of the molecule and produce a hemolytic anemia. One of the most common defects is the substitution of valine for glutamic acid in the sixth position of the β chain, producing hemoglobin S. This substitution causes a profound change in the solubility characteristics of the hemoglobin. When deoxygenated, hemoglobin S polymerizes to form rigid, rodlike intracellular aggregates, which deform the cell and produce the typical sickle cell shape seen in sickle cell anemia (see pages 52 and 53).

A quantitative defect in hemoglobin synthesis results from deficient iron supply or an impairment of porphyrin or globin production during cell development.[8] In these situations, microcytic, hypochromic red cells appear in circulation. The reduction in red cell volume demonstrates the key role played by hemoglobin content in determining cell size.

Erythrocyte Life Cycle and Breakdown

The normal red cell has a finite life span of about 120 ± 20 days. While in circulation, it is subjected to a great variety of metabolic and mechanical stresses. Turbulence of blood flow or damage to the endothelial lining of arterial vessels with deposition of fibrin can produce red cell fragmentation and intravascular hemolysis. When a strong physical force, such as the pounding of feet on the ground, is applied to a vascular area, red cells can be ruptured in circulation. As cells become older, certain catabolic changes occur, leading to a decrease in cell flexibility. This decrease makes it more difficult for the red cell to traverse the microvasculature and, at some point, results in either lysis of the cell in circulation or phagocytosis and removal by the reticuloendothelial system.

Although all reticuloendothelial cells participate in the destruction of senescent red cells, those in the spleen are anatomically the most sensitive detectors of any red cell abnormality (Fig. 14). Blood enters the reticular network of the splenic pulp through terminal arterioles. Flow through the red cell pulp is slow, and the volume of plasma is reduced, further stressing the metabolic machinery of the red cell. In order to reenter the venous circulation, the red cell must squeeze through a small, 2- to 5-μm orifice in the sinusoidal wall. This is the ultimate test for the pliability of the erythrocyte. Rigid cells are trapped and phagocytized by the splenic reticuloendothelial cells. Abnormal cellular inclusions are detected and removed. The quality control exerted by the spleen on the circulating red cell mass is most evident in the blood of splenectomized individuals.[22] In this blood, one may see a spectrum of red cell abnormalities including red cells containing nuclear remnants (How-

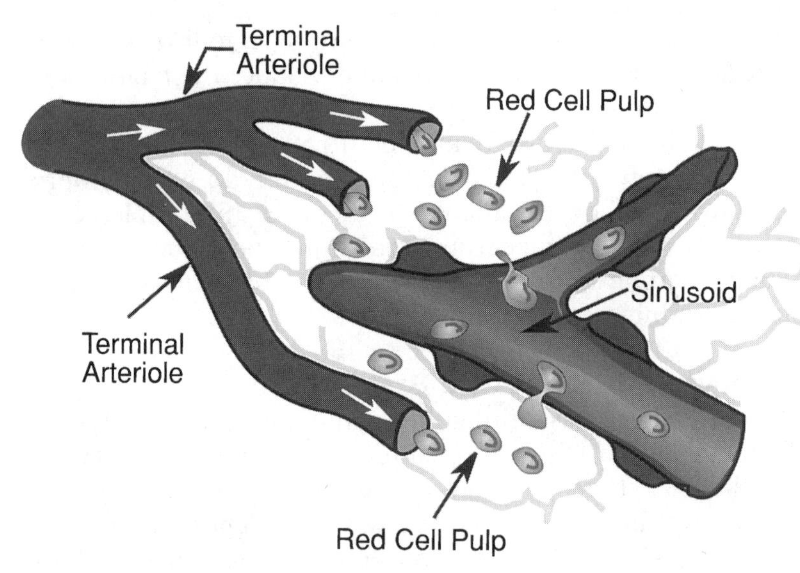

Figure 14. Structure and function of the spleen, which is responsible for detecting red cell abnormalities and removing such cells from circulation. The unique structure of the spleen provides a severe test of cell viability. Red cells are delivered to the spleen through terminal arterioles and slowly perfuse the red cell pulp. During this time, the volume of plasma is reduced, thereby stressing the cell's metabolic machinery. The red cell must then pass through a small, 2- to 5-μm orifice to reach the splenic sinusoid and escape back into the general circulation. Intracellular inclusion bodies or abnormally rigid cells are trapped by this filtering process and phagocytized by the sinusoidal reticuloendothelial cells.

ell-Jolly bodies), denatured hemoglobin inclusions (Heinzbocher bodies), iron granules (siderocytes), and a number of misshapen or fragmented cells (target cells, schistocytes, and tear drops).

In addition to their mechanical testing of cell pliability, reticuloendothelial cells can recognize antibody globulin on the red cell surface. Splenic reticuloendothelial cells have receptors for the Fc fragment of immunoglobulins and can remove and destroy red cells coated with IgG. In addition, both liver and spleen reticuloendothelial cells recognize the C3b component of complement on the surface of a cell and, even in the absence of a pliability change, trap and phagocytize the cell.

Red cell removal by the reticuloendothelial system is referred to as *extravascular destruction.* It provides the most efficient method for processing senescent cells with recovery of essential components such as amino acids and iron.[23] Within the reticuloendothelial cell, the red cell is attacked by lysosomal enzymes. The membrane is disrupted and the hemoglobin molecule broken down by an enzyme, heme oxygenase (Fig. 15). The freed iron is either returned to plasma transferrin and transported back to the erythroid marrow or stored within the reticuloendothelial cell as ferritin and hemosiderin. Amino acids are redirected to the body protein pool. The protoporphyrin ring of heme is broken at the α-methene bridge and its α-carbon exhaled as carbon monoxide. The opened tetrapyrrole (bilirubin) is carried by plasma albumin to the liver, where it is conjugated to form the glucuronide and excreted in the bile. The bilirubin glucuronide entering the gut is then converted through bacterial action to urobilinogen. Most of the urobilinogen is excreted in the stool as urobilin, the orange-yellow compound that gives stool its color. Approximately 10% to 20% of the urobilinogen is reabsorbed unchanged and either excreted in urine or returned to the gut via an enterohepatic cycle. In the presence of liver disease, the enterohepatic cycle is impaired and a greater proportion is excreted in the urine.

Although most senescent red cells undergo extravascular destruction, some cells break down in circulation. This *intravascular destruction* normally accounts for less than 10% of red cell loss, although it can increase significantly in certain disease states (Fig. 16).[23] The fate of hemoglobin released directly into the bloodstream differs from that observed in extravascular destruction. The free hemoglobin tetramer is unstable in plasma and undergoes dissociation into α-β dimers, which are quickly bound to the plasma protein

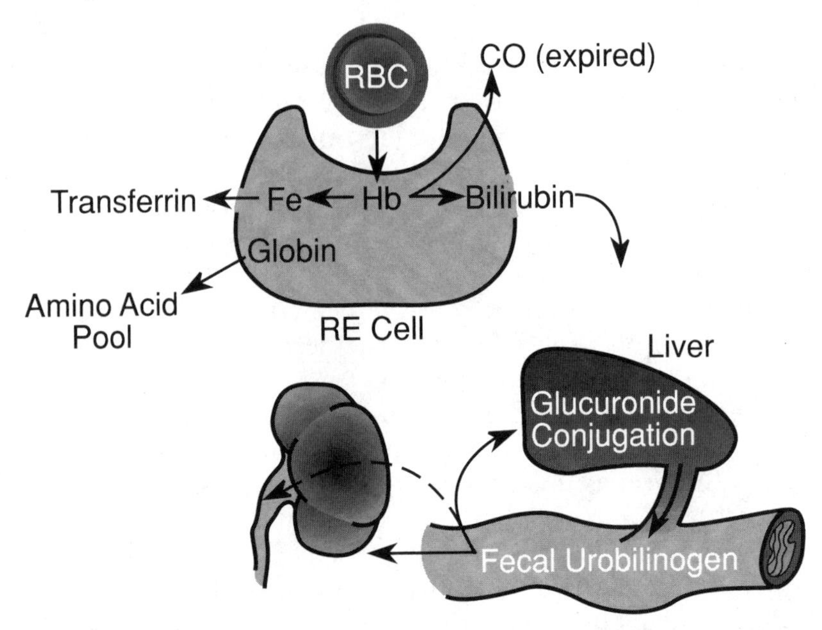

Figure 15. Red cell destruction by the reticuloendothelial system. Normally, senescent red cells are phagocytized by reticuloendothelial cells and the hemoglobin broken down into its essential components. The recovered iron is returned to transferrin for new red cell production, and the amino acids from the globin portion of the molecule are returned to the general amino acid pool. The protoporphyrin ring of heme is broken at the α-methene bridge and its α-carbon exhaled as carbon monoxide. The remaining tetrapyrole leaves the reticuloendothelial cell as indirect bilirubin and travels to the liver, where it is conjugated for excretion in the bile. In the gut, the bilirubin glucuronide is converted to urobilinogen for excretion in stool and urine.

haptoglobin. Formation of the haptoglobin-hemoglobin complex prevents renal excretion of plasma hemoglobin and stabilizes the heme-globin bond. The complex can then be removed from circulation by hepatocytes and is processed within these cells in a fashion similar to that described for the reticuloendothelial cells. There is a limit to the capacity of the haptoglobin-binding mechanism, and a sudden intravascular release of several grams of hemoglobin can exceed the binding capacity. Furthermore, because haptoglobin itself is removed from circulation as the hemoglobin-haptoglobin complex is catabolized, a decrease in or absence of haptoglobin may be used to indicate increased intravascular hemolysis.

Once haptoglobin is depleted, unbound hemoglobin dimers are free to be filtered by the renal glomerulus, where they are ei-

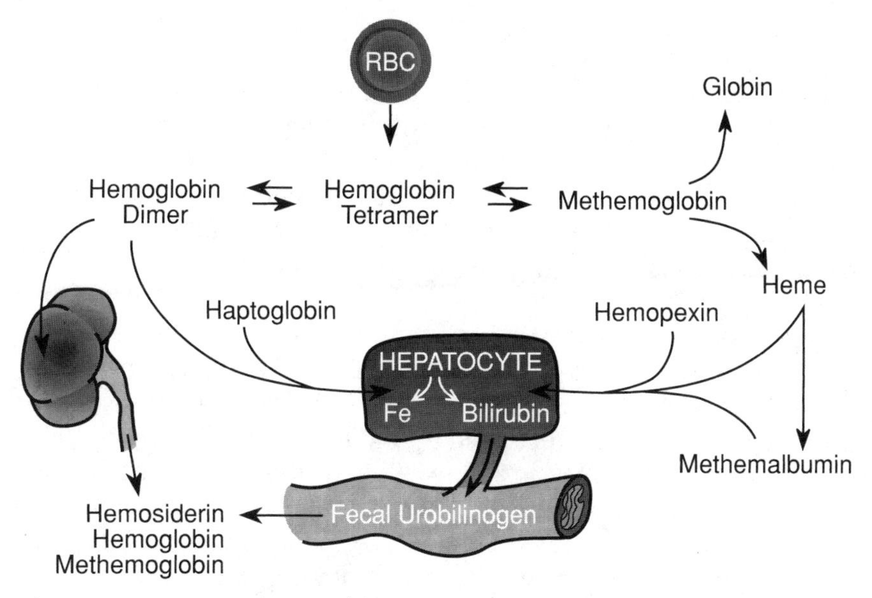

Figure 16. Intravascular red cell hemolysis. Red cells can also undergo intravascular hemolysis with the release of hemoglobin into circulation. The free hemoglobin tetramer is unstable and rapidly dissociates into α-β dimers, which bind to haptoglobin and are removed by the liver. Hemoglobin may also be oxidized to methemoglobin and dissociate into its globin and heme moieties. To a limited extent, free heme can be bound by hemopexin and/or albumin for subsequent clearance by hepatocytes. Both pathways help recover heme iron for the support of hematopoiesis. Once haptoglobin is depleted, unbound hemoglobin dimers are excreted by the kidney as free hemoglobin, methemoglobin, or hemosiderin.

ther resorbed by tubular cells and converted to hemosiderin or, when the tubular uptake capacity is exceeded, excreted as free hemoglobin or methemoglobin in the urine.[24] As much as 5 g per day of filtered hemoglobin may be reabsorbed without exceeding the tubular uptake capacity. However, most of the hemoglobin iron trapped within the tubular cells as hemosiderin is ultimately lost when the tubular cells are desquamated. Thus, renal processing of filtered hemoglobin is accompanied by the excretion of either hemosiderin alone, a combination of hemosiderin and hemoglobin, or, in the case of an acute hemolytic event, only hemoglobin. With chronic intravascular hemolysis, sufficient iron can be lost as hemosiderin to produce an iron deficiency state.

The plasma hemoglobin that is not bound to haptoglobin may be oxidized to methemoglobin, whereupon heme groups dissociate and are bound to another transport protein, hemopexin.

Heme-hemopexin complexes are also cleared from circulation by hepatocytes and then catabolized. Like haptoglobin, hemopexin may become exhausted when the amount of plasma heme exceeds the plasma hemopexin concentration. In that event, excess heme groups combine with albumin to form methemalbumin, which circulates until more hemopexin is formed to initiate transfer to the liver. Because this may take several days, methemalbumin serves as a marker for a prior intravascular hemolytic event.

Erythropoietin Regulation of Erythropoiesis

The erythron responds to sustained imbalances of oxygen supply and tissue oxygen demand by increasing hemoglobin concentration. Regulation is accomplished by erythropoietin, a glycoprotein hormone produced by the kidney.[25] Peritubular interstitial cells within the kidney are capable of detecting a decrease in available oxygen, whether it is caused by a decrease in hemoglobin, a decrease in the oxygen content of hemoglobin, or an increase in hemoglobin affinity for oxygen. In other words, these cells respond to a decrease in the amount of oxygen that can be extracted from arterial blood at a given oxygen tension. One unique feature is a relative insensitivity to changes in blood flow. This occurs because renal oxygen consumption and changes in blood flow are somehow balanced, so that regulation, which expresses the relationship between oxygen supply and tissue requirement, is unaffected.

Increased production of erythropoietin by the kidney is associated with a recruitment of existing interstitial cells to initiate the transcription of EPO mRNA from a single copy of a gene on the long arm of chromosome 7. The final product, erythropoietin, is a glycoprotein of 30,000 d, 60% of which is a single polypeptide chain with 165 amino acids and 40% of which is carbohydrate (Fig. 17). Normally, only picomolar quantities are present in blood, indicating the extreme potency of the molecule. By radioimmunoassay, the basal plasma level of erythropoietin in human beings is 9 to 26 mU/mL. Within a matter of hours after the onset of an anemic or hypoxic stimulus, a rising tide of erythropoietin can be detected in plasma, with the response correlating with the severity of the stimulus (Fig. 18).

Erythropoietin stimulates committed stem cells (BFU–E/CFU–E) to increase the number of erythroid precursors and, eventually, the number of red cells entering circulation. The hormone binds to specific receptors on the surface of CFU–E and is rapidly inter-

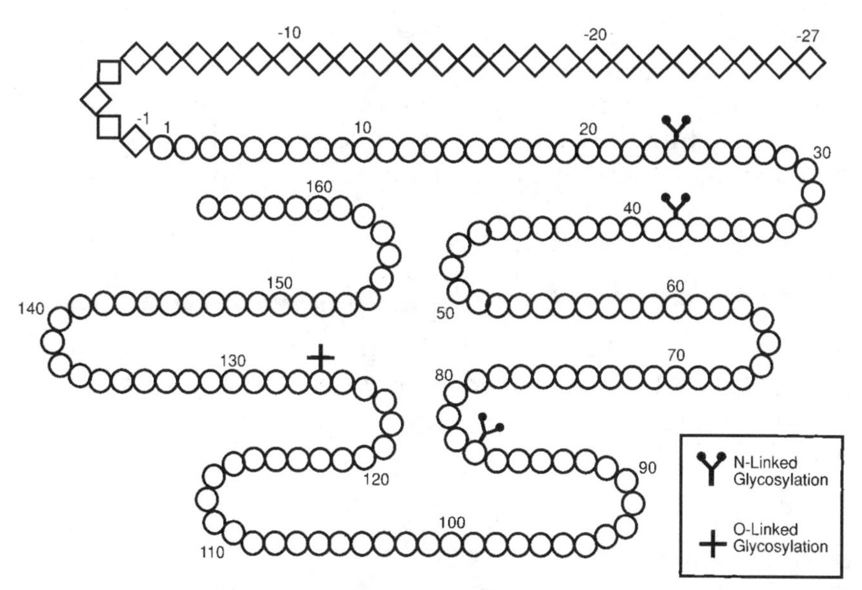

Figure 17. Amino acid sequence of erythropoietin. Erythropoietin is a single polypeptide chain of 165 amino acids (*open circles*). It is a heavily glycosylated molecule, which determines the clearance characteristics from circulation. The 27 amino acids shown as squares represent a single peptide that is cleaved after the protein is secreted by the manufacturing cell.

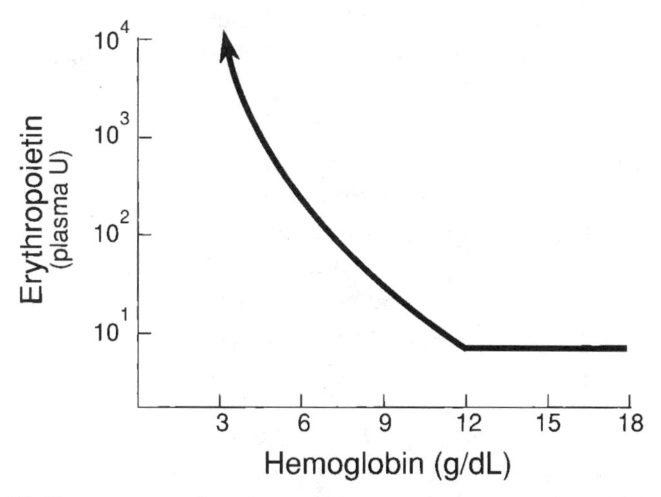

Figure 18. The response of erythropoietin secretion to anemia. With worsening anemia, there is little change in the plasma erythropoietin level until the hemoglobin level falls below 12 g/dL. Subsequently, the plasma and urine erythropoietin levels increase logarithmically. As long as renal function is normal, marrow function intact, and iron supply adequate, the rise in erythropoietin and the magnitude of the erythroid marrow response correlate with the severity of the anemia.

nalized and degraded. The earliest detectable cellular action is an immediate increase in intracellular calcium ion, followed in 1 to 2 hours by transcription of mRNA for several erythroid proteins, including the globin polypeptide chains. Increased DNA synthesis is detected by 20 to 40 hours. The full proliferative response of the erythroid marrow requires several days but does not require a continued high titer of erythropoietin. This fact is recognized in planning therapy with intravenous or subcutaneous recombinant human erythropoietin, with which a dosage schedule of three times a week is sufficient to give a sustained proliferative response. The very important role of iron supply in determining the response to erythropoietin is discussed on page 35.

Oxygen Supply

The red cell, although essential for oxygen transport, is but one component in the oxygen supply system. A number of other physiologic components must also function in a highly integrated fashion (Fig. 19).[26] Pulmonary function and hemodynamic factors such as cardiac output, regional blood flow, blood volume, and blood viscosity are all major components. Each has its own behavior, and the compensation by each for hypoxia varies according to the nature of the physiologic stress.

The Red Cell in Oxygen Transport

Red cell oxygen supply is determined by the number of erythrocytes and their hemoglobin content as well as the capacity of the hemoglobin to release oxygen. At a hemoglobin concentration of 15 g/dL, every 100 mL of whole blood carries approximately 20 mL of oxygen (1.3 mL O_2/g of hemoglobin). Arterial blood enters tissues at an oxygen tension of 95 torr and venous blood exits with an average oxygen tension of 40 torr. Thus, under basal conditions, approximately 20% to 25% of the oxygen carried by red cells is released, and a corresponding amount of carbon dioxide is accepted for transport back to the lungs. The capacity of the transport system is reduced by a decrease in circulating red cells or hemoglobin. The effect of an increase in circulating red cells above normal is minimal because of the increase in blood viscosity that this produces.

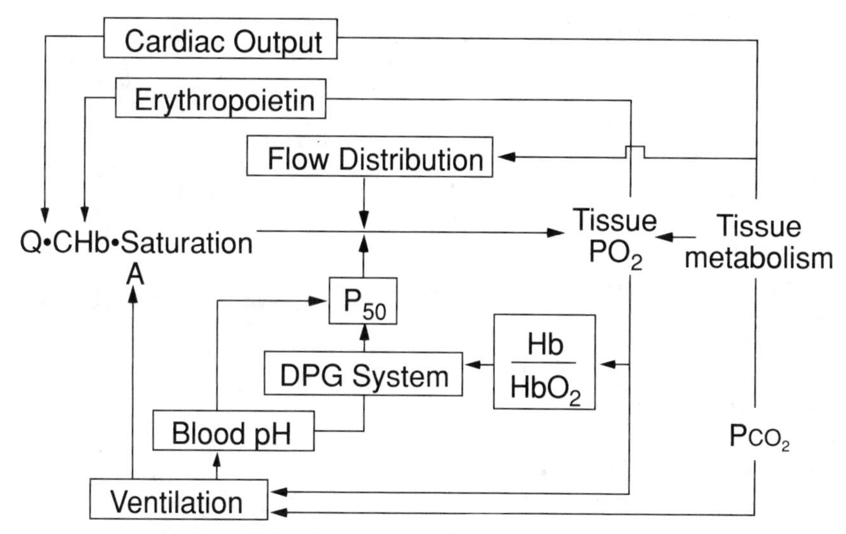

Figure 19. The relationship of various physiologic components to tissue oxygen supply. The tissue partial pressure of oxygen (Po_2) is a product of the cardiac output, hemoglobin concentration, and the oxygen saturation of arterial blood. In addition, the distribution of blood flow to individual tissues and shifts in the hemoglobin-oxygen dissociation curve adjust oxygen delivery, especially when local tissue metabolism demands a greater oxygen supply. Each of the major components may be modified by oxygen need. A general increase in tissue metabolism stimulates cardiac output, increases ventilation, and, by its influence on blood pH, shifts the hemoglobin-oxygen dissociation curve to release more oxygen to tissues. Sustained reductions in the tissue Po_2 not only stimulate an increased production of red cells by erythropoietin, but also increase minute ventilation and stimulate the intracellular production of 2,3-DPG to shift the hemoglobin-oxygen dissociation curve for greater oxygen delivery.

Oxygen delivery can also be influenced by altering the affinity of hemoglobin for oxygen. Hemoglobin-oxygen affinity is determined by a number of factors, such as temperature, pH and carbon dioxide concentration, and the red cell organic phosphate content. At a pH of 7.4, a CO_2 tension of 40 torr, and a temperature of 37° C, 50% of the oxygen carried by hemoglobin is released when the Po_2 is 27 torr. This point is referred to as the P_{50} of the hemoglobin-oxygen dissociation curve (Fig. 20).

In vivo, the position of the curve and its P_{50} are affected by tissue metabolism.[27] For example, exercising muscle generates acid metabolites, thus shifting the dissociation curve to the right and increasing the P_{50} level of red cells as they pass through its capillaries. This permits more oxygen to be released for any tissue Po_2. If

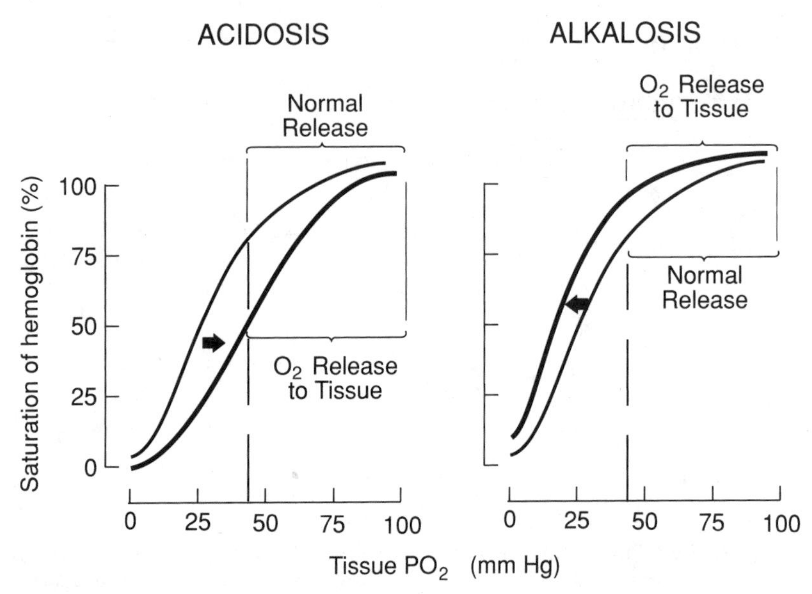

Figure 20. The Bohr effect—the relationship of pH and carbon dioxide concentration to hemoglobin-oxygen affinity. The position of the hemoglobin-oxygen dissociation curve and its P_{50} are affected by tissue metabolism and blood pH. When increased tissue metabolism releases sufficient acid products to produce acidosis, the curve is shifted to the right, permitting more oxygen to be released. As shown, even a small shift in the curve significantly increases the amount of oxygen released to the tissue. This Bohr effect occurs instantaneously at the site of tissue metabolism and, therefore, can effectively increase oxygen delivery to a localized area. In contrast, an increase in pH, usually occurring as a generalized alkalosis, shifts the hemoglobin-oxygen dissociation curve to the left, reducing the amount of oxygen released to tissues.

there is a simultaneous local fall in tissue oxygen tension, the amount of oxygen released is even greater. In the case of working muscle, 75% or more of the hemoglobin-bound oxygen in red cells traversing the tissue can be released as a result of the Bohr effect and a reduction in tissue Po_2 levels. An increase in pH has the opposite effect. For example, the alkalosis of hyperventilation shifts the hemoglobin-oxygen dissociation curve to the left, thereby reducing oxygen availability to tissues.

In addition to these environmental effects, the red cell is itself able to alter the affinity of its hemoglobin. When oxygen supply is inadequate to meet tissue needs, tissue oxygen tension falls and the arterial-venous difference in oxyhemoglobin concentration widens. The resultant rise in deoxyhemoglobin within the red cell stimulates increased production of 2,3-DPG, regardless of whether the change is caused by arterial oxygen desaturation, cardiac failure, or anemia.[26] This increased 2,3-DPG production lowers the affinity of hemoglobin for oxygen, as reflected in a sustained right shift of the affinity curve. Unlike the immediate change resulting from the Bohr effect, this adaptation takes 12 to 36 hours. Once accomplished, however, the resultant increase in P_{50} level permits a greater delivery of oxygen at a more normal tissue oxygen tension. For the DPG response to occur, there must be an adequate supply of inorganic phosphate in the plasma.

Erythrocyte DPG metabolism is also affected by generalized and sustained acidosis or alkalosis, resulting in a return of the in vivo P_{50} level to normal. For example, when a diabetic patient becomes acidotic, the red cell dissociation curve initially shifts to the right because of the Bohr effect. However, if the acidosis continues for more than a few hours, the rate of red cell glycolysis decreases, resulting in a fall in the red cell 2,3-DPG level. The affinity curve then shifts back to a normal position; that is, the Bohr effect is counterbalanced by the reduction in the 2,3-DPG level. This change in oxygen delivery of the red cells is physiologically appropriate for the needs of tissue; its only disadvantage is the disparity in the time required for the DPG response as compared to the Bohr effect. This problem becomes evident when the acidosis is rapidly reversed by therapy. With the sudden loss of the Bohr effect, the hemoglobin-oxygen dissociation curve shifts to the left because the cell is still depleted of intracellular 2,3-DPG. Reconstitution of the 2,3-DPG and return to a normal P_{50} level may require several days, and, again, require an adequate supply of inorganic phosphate.

Pulmonary Function and Hemodynamic Factors

Tissue oxygen supply also requires efficient loading of oxygen onto the hemoglobin of red cells as they pass through the lungs. The shape of the hemoglobin-oxygen dissociation curve favors the loading of oxygen, and more than 95% saturation is achieved at the normal alveolar Po_2 level of 100 mm Hg. Because of the flat slope at loading oxygen tensions, hemoglobin oxygen loading in the lungs is relatively insensitive to changes in the position of the curve in contrast to the marked effects of curve shifts on oxygen release.

Ventilation is regulated primarily by the carotid body, which is sensitive to both carbon dioxide and oxygen tensions. The normal individual responds to a fall in alveolar PO_2 and arterial oxygen saturation with an increase in minute ventilation.[27] This is an important compensation for the individual at high altitude. However, ventilation is insensitive to hemoglobin concentration because the extremely high blood flow through the carotid body prevents local changes in oxygen tension. Therefore, neither anemia nor cardiac failure appreciably affects minute ventilation of the resting individual.

Redistribution of blood flow can act as a major defense against tissue hypoxia. Cellular metabolism and the production of acid metabolites such as carbon dioxide and hydrogen ion promote an increased regional blood flow and oxygen supply. Through this local mechanism, individual organs modify blood flow according to their needs. This modification may occur without a change in cardiac output because of reciprocal reductions in flow to other tissues. In effect, a trade is made between tissues that require more oxygen and tissues having excessive oxygen supply, such as the kidney and skin. This trade-off is accomplished through sympathetic nervous system activity and catecholamine release. Reductions of as much as 50% in oxygen supply caused by cardiac failure or anemia can be compensated for by shifts in regional blood flow and a simultaneous change in hemoglobin-oxygen affinity.[28] Such adjustments in themselves are not without a price, namely, a reduction in renal excretory capacity and impaired heat regulation by the skin.

Cardiac output is little affected by arterial oxygen desaturation, and anemia does not directly affect resting cardiac output until the hemoglobin falls to less than 8 g/dL. Cardiac output responds primarily to increases in total oxygen consumption, as with physical activity or such hypermetabolic states as thyrotoxicosis and pregnancy.

Sufficient intravascular fluid and an appropriate vascular tone are also required to maintain an active circulation between lungs and body tissues. Adjustments for variations in blood volume caused by changes in red cell mass must be made if cardiac output and regional blood flow are to be maintained. With the gradual development of anemia, a rise in plasma volume is accomplished by increased saline retention and increased albumin synthesis. This balance between blood volume and the vascular container can be lost in situations of acute blood loss or hypovolemia resulting from severe saline depletion (vomiting, diarrhea, or excessive sweating), a selected depletion of the intravascular compartment as a result of the loss of oncotic protein (anaphylaxis, burns, impaired albumin production, or renal loss) or marked vasodilatation (reflex vasodilatation or sepsis).

Red Cell Production and Destruction

Measurements

The capacity of the erythron to respond to an anemic or hypoxic stimulus can be characterized by a variety of measurements of red cell production and destruction (Fig. 21).[29] Often measurements

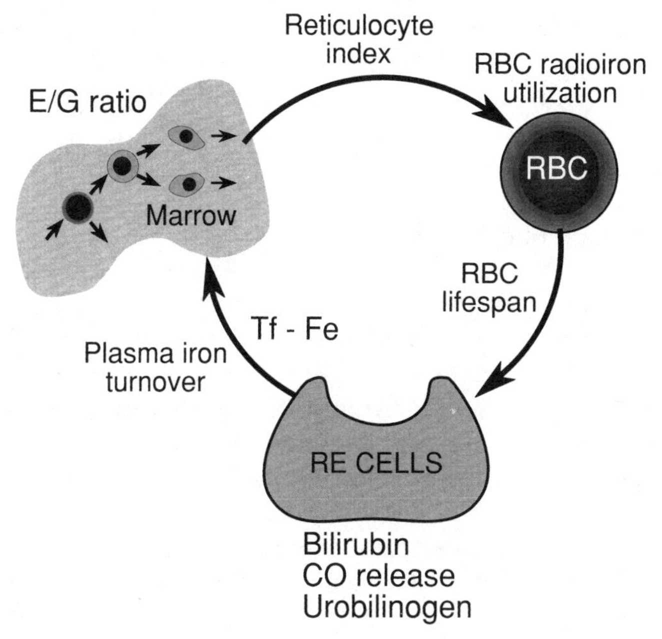

Figure 21. A variety of measurements are available to characterize erythroid marrow production and red cell destruction. In terms of the erythroid marrow, cell culture techniques can be used to define relative numbers of stem cells. Plasma iron turnover (PIT) studies permit accurate measurement of total erythroid marrow production and effective production of adult red cells (RBC radioiron use). From a clinical standpoint, the marrow E:G ratio for total production and the reticulocyte index for effective red cell production can serve the same purpose. Red cell destruction rates may be evaluated using techniques such as the ^{51}Cr red cell life span, bilirubin turnover, carbon monoxide release, and urobilinogen turnover.

show a combined decrease in production with some increased destruction, in which case the anemia is identified with the component of greatest magnitude.

The three elements in red cell production to be considered are the erythropoietin drive, the capacity of the stem cell population to respond, and the integrity of the maturation process. Plasma erythropoietin may be measured by a radioimmunoassay; however, it is not practical clinically, and its interpretation depends on a number of factors. A range of values is found in anemic patients, depending on the severity and duration of the anemia, renal function, compensatory changes in hemoglobin oxygen delivery, and the degree of marrow proliferation. The relative number of stem cells can also be examined using the technique of cell culture, but this examination is also of little clinical use.

More quantitative indicators of erythroid marrow response are found in the visible marrow. The most practical marrow measurement is the relative number of nucleated erythroid versus granulocyte precursors in the marrow aspirate, that is, the E:G ratio. This estimate is valid only if the granulocyte marrow is normal. There are other measurements of total erythroid marrow activity that, although not in general clinical use, are more precise and have done much to characterize erythropoiesis. One of these is radioiron kinetics.[30] The plasma iron turnover (PIT) converted to a corrected erythron iron turnover (EIT) can be used to quantitate the number of erythroid transferrin receptors. This bears a direct relationship to the number of erythroid precursors and indicates the degree of erythroid proliferation. A secondary measurement in close agreement with the ferrokinetic measurement is the determination of plasma transferrin receptors.[31] Most plasma transferrin receptors are derived from maturing red cells, so that the plasma receptor level also reflects the number of erythroid precursors. However, plasma transferrin receptor measurements are valid only if the iron supply is adequate. With iron deficiency, there is an increase in the number of transferrin receptors per cell that results in overestimation of the erythroid cell population.

Effective red cell production (i.e., the actual number of red cells entering circulation) can be measured from the incorporation of a tracer dose of radioiron into circulating red cells. A more practical measurement (and the one in general use) is to determine the number of reticulocytes entering circulation.[32] For a true estimate of effective production, the reticulocyte count (the percentage of reticulocytes in circulation) must be corrected for both the degree of anemia and the effect of increased erythropoietin stimulation (see page 34). This provides a reticulocyte production index that, although only an approximate measure of red cell production, is quite satisfactory for clinical diagnosis.

Measurements of red cell production can be interpreted only in light of some estimate of red cell destruction. Several means of evaluating overall hemoglobin catabolism and red cell survival are available. They range from simple screening tests to highly quantitative measurements and from those reflecting total breakdown (marrow and blood) to those reflecting only the breakdown of circulating cells. Total breakdown can be determined from the amount of exhaled carbon monoxide (derived from the opening of the protoporphyrin ring) or the amount of urobilinogen in stools.[33] Neither of these measurements is very practical clinically. Therefore, plasma bilirubin and lactate dehydrogenase (LDH) levels are used as semiquantitative indicators of total breakdown.[34] A direct measurement of red cell survival using isotope-tagged erythrocytes can provide a more quantitative measurement of circulating red cell destruction.[35]

The relationship of these various measurements of red cell production and destruction is summarized in Table 1. To apply such measurements in the clinical diagnosis of anemia, it is important to understand the usefulness of each in defining total versus effective red cell production and destruction. In the clinical setting, the marrow E:G ratio is the best measurement of total red cell production, whereas the reticulocyte index is used to define effective red cell production. Levels of indirect bilirubin and LDH are readily available and provide useful information on total red cell destruction.

Table 1 • **Quantitative Measurements of Red Blood Cell Turnover**

RBC Production	RBC Destruction
Total	Total
Marrow E:G ratio	CO excretion
Erythron iron turnover	Stool urobilinogen
	Bilirubin turnover
Effective	Effective
Reticulocyte index	^{51}Cr survival
Radioiron reappearance	

Stimulated Erythropoiesis

Proper interpretation of any of these measurements requires further definition of the normal behavior of the erythron. This analysis must take into account not only the characteristics of erythropoiesis under basal conditions but also the increased erythropoietic response expected with anemia (Table 2). In the basal state, the normal man has a circulating red cell mass of approximately 2000 mL (750 g of hemoglobin), which turns over (that is, is produced and destroyed) at a rate of approximately 1% per day because normal red cells survive for 100 to 120 days. The number of red cell precursors in the marrow is less than one-tenth of the number of red cells in circulation, and most of these are marrow reticulocytes. Red cell size and morphology are uniform, and circulating reticulocytes as well as measurements of pigment turnover are at basal levels.

These "normal" quantities, by themselves, do not define the productive capacity of the erythroid marrow. They represent normal values for the basal state when oxygen supply is adequate, red cell production is effective, and red cell survival is normal. Under the stimulus of hypoxia or anemia, the marrow is able to increase its level of production several times. This proliferative capacity is as much a "normal" characteristic of the anemic state as the basal pa-

Table 2 • **Basal and Stimulated Erythropoiesis***

	Basal (Hb 15 g/dL)	Stimulated (Hb 8 g/dL)	
		Acute	*Chronic*
Marrow (cell number × 10^9/kg)			
Nucleated red cells	4.5	13.5	27
Reticulocytes	7.5	12	28†
Blood (cell number × 10^9/kg)			
Reticulocytes	3.0	18	36†
Adult red cells	300	150	150
Reticulocyte index	1.0	3	6

*Assumes adequate iron supply.
†Reflects shift of marrow reticulocytes into the circulation.

rameters are of the nonanemic state. To define a "normal" capacity of the erythron to respond to anemia, four factors must be taken into account. First, there is the time lag of several days to a week before the full proliferative response of the marrow is observed. This is the time required for red cell precursors to differentiate, complete four divisions, and mature. Although a visible proliferation of marrow erythroid precursors is observed within 2 to 3 days of the onset of anemia or hypoxia, a full rise in the reticulocyte index does not occur for several more days.

Second, the magnitude of the response is affected by the degree of erythropoietin stimulation, which is usually related to the severity of the anemia.[25] Moderate reductions in the hemoglobin level, which are readily compensated for by a shift in hemoglobin-oxygen affinity, do not generate a maximal marrow response (see Fig. 19). At hemoglobin levels of 10 to 11 g/dL or higher, the marrow production response generally does not exceed two times basal rate. An apparent exception to this rule is seen in patients with certain hereditary hemolytic anemias, in whom high levels of erythropoiesis may be maintained at near normal hemoglobin levels. Some of these patients show an impaired release of oxygen from their hemoglobin that would result in a higher level of erythropoietic stimulation, but the important factor is the enhanced iron supply associated with red cell hemolysis.

The third and most important factor in determining the normal response is iron.[36] If otherwise normal people experience an increased rate of red cell destruction resulting in a fall of hemoglobin to below 10 to 11 g/dL, most will increase red cell production to levels between three and five times basal rate by 1 week. This is the response that is expected of patients with hemolytic anemias. In contrast, if otherwise normal people are rapidly bled to a hemoglobin below 10 g/dL, rates of red cell production vary from little or no response to an increase of two to three times basal rate. This difference in marrow response reflects variations in iron supply. Some individuals have virtually no iron stores, and their erythropoietin-stimulated marrow cannot produce more red cells because of the limitation in iron supply. Other individuals have modest iron stores and can increase erythropoiesis to two to three times basal rate. Those with abundant stores show a near-optimal response of three to four times basal. Few, if any, can match the five times or higher levels seen in the hemolytic anemia patients who have larger amounts of iron available, corresponding to the daily red cell breakdown.

The fourth factor is chronicity. When increased rates of red cell destruction are sustained over long periods of time, marrow production increases still further, to levels in excess of six to eight times basal. This requires not only an excellent iron supply, provided by the catabolism of red cells, but also an expansion of the erythroid marrow into long bones.

Abnormal Erythropoiesis

The capacity of the normal marrow to increase its red cell production is an essential consideration in the differential diagnosis of severe anemia. Along with red cell destruction, it permits one to classify red cell disorders according to whether there is a defect in proliferation (hypoproliferative anemia), maturation (ineffective erythropoiesis), or red cell survival (hemolysis).

These patterns of marrow dysfunction can be illustrated using ferrokinetic measurements (Fig. 22).[37] In the normal individual, intravenously injected radioiron immediately binds to transferrin and then disappears from circulation with a half-time of 1 to 2 hours. The destination of most of this iron is the erythroid marrow. After a transit time of 3 to 4 days, which represents the maturation time of the red cell precursors, most of the radioactivity leaves the marrow and enters the circulation as a part of the newly formed hemoglobin and reticulocytes. A small amount of activity also appears in the liver and spleen.

When compared to this normal ferrokinetic pattern, the patient with a severe defect in marrow proliferation (e.g., aplastic anemia) has a prolonged clearance of iron from the plasma of 3 to 5 hours. At the same time, little activity occurs in the marrow, but a much greater uptake occurs in the liver. In effect, this is the pattern one would expect when there are few red cell precursors in the marrow to accept transferrin iron.

In the patient with a maturation abnormality of red cell precursors, radioiron is cleared rapidly from the blood with a normal reciprocal uptake by the marrow. However, the subsequent incorporation of the iron into hemoglobin and its appearance in circulating reticulocytes are markedly reduced. Instead, defective cell marrow precursors are destroyed during maturation or immediately after their appearance in circulation, and their radioiron is found in the reticuloendothelial cells of the bone marrow, spleen, and liver. This pattern of defective maturation is referred to as *ineffective erythropoiesis.*

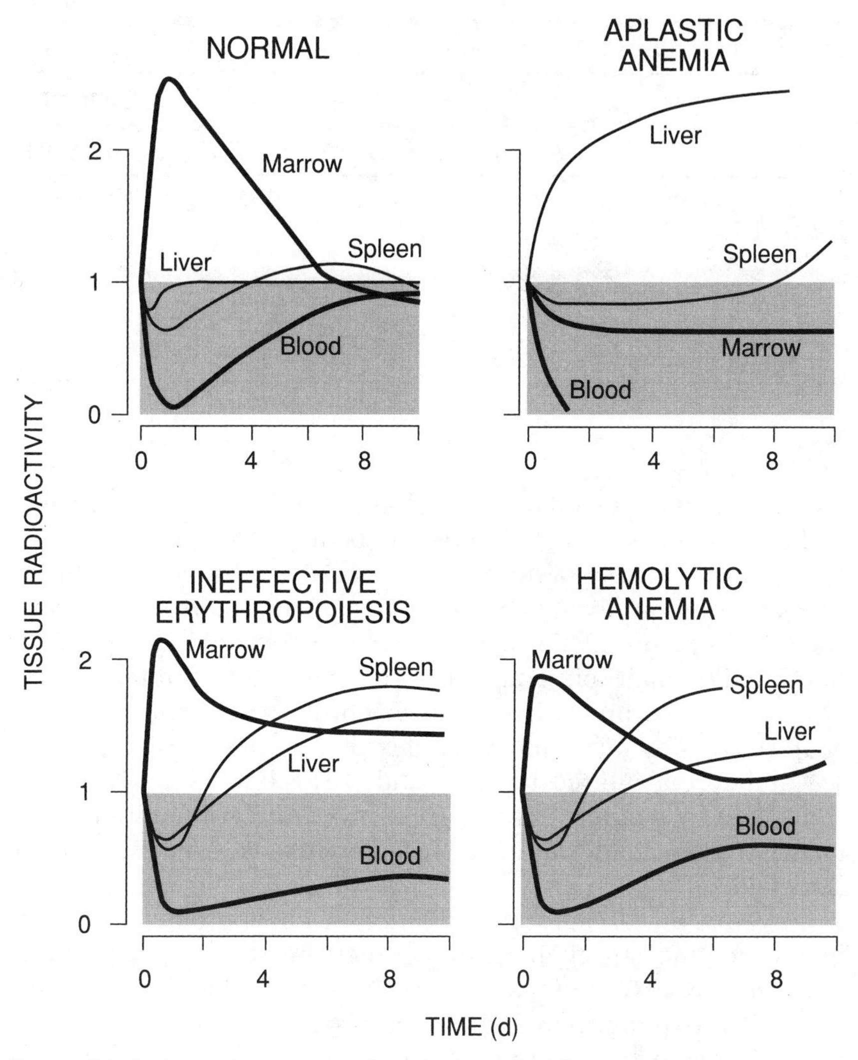

Figure 22. Patterns of marrow dysfunction as defined by their ferrokinetic profiles (see text).

In patients with hemolytic anemia (i.e., increased destruction of circulating red cells) the same accelerated clearance of radioiron from the blood with localization in the marrow is seen. However, there is a greater early rise in circulating red cell activity because the newly formed red cells survive for a time in circulation. The appearance of radioactivity in circulation may be short-lived if the destructive process involves newly formed red cells. In that event, the

Table 3 • **Patterns of Abnormal Erythropoiesis**

	Production		Destruction
	Retic Index	*Marrow E:G Ratio*	*Bilirubin*
Normal (basal)	1.0	1:3	Normal
Anemia			
Hemolytic	>3.0	>1:1	Increased
Hypoproliferative	<2.0	<1:2	Decreased
Ineffective erythropoiesis	<2.0	>1:1	Increased

level of activity in blood begins to decline within the first few days, and the isotope is relocated to reticuloendothelial cells.

Clinically, these patterns of abnormal erythropoiesis can be defined using more available measurements of marrow E:G ratio (total red cell production), reticulocyte index (effective production), and the bilirubin and LDH levels (indicators of total red cell destruction). As previously defined, normal *basal erythropoiesis* is characterized by a reticulocyte index of 1, a marrow E:G ratio of 1:3, and normal indirect bilirubin and LDH values (Table 3). When stimulated by a moderately severe anemia, the normal individual increases both total and effective production at least twofold to threefold.

Against this background, patients with hypoproliferative anemia are defined by their failure to increase red cell production. Their marrow E:G ratio is less than 1:2 and their reticulocyte index is less than two times normal. Because red cell production has not increased to compensate for the reduction in circulating red cell mass, overall red cell turnover is below normal, resulting in an indirect bilirubin count less than 0.4 mg/dL and a normal LDH value. In contrast, a patient with a maturation disorder will show a high marrow E:G ratio, indicating red cell precursor proliferation, but a low reticulocyte index. The increased red cell turnover associated with this ineffective erythropoiesis results in an indirect bilirubin count greater than 0.7 mg/dL, and LDH values often in excess of 1000 IU/L. Finally, substantial increases in the marrow E:G ratio, reticulocyte index, and measurements of destruction (the indirect bilirubin and LDH levels) are typical of the hemolytic anemias.

Chapter

2 The Detection of Anemia

Signs and Symptoms

In the otherwise healthy individual, mild anemia is usually asymptomatic, although it may be associated with a slight increase in dyspnea, palpitations, and sweating with exercise.[38] As anemia becomes more severe, these symptoms are more pronounced, and work capacity is reduced. Severe anemia is associated with marked dyspnea, palpitations, and a pounding headache with exertion.[39] Other symptoms and signs that appear to have a physiologic basis include 1) sodium retention sufficient to produce dependent edema; 2) loss of appetite and indigestion secondary to limitations in intestinal oxygen supply; 3) generalized weakness, dizziness, and occasional syncope related to compromised vasomotor regulation; and 4) hypoxic symptoms of the central nervous system including insomnia, inability to concentrate, and even disorientation. The severity and nature of these symptoms vary according to the age and general health of the patient. Previously asymptomatic vascular disease in the presence of anemia can be manifested by localized ischemic manifestations caused by the combination of decreased blood flow and reduced blood oxygen content. Consequently, elderly patients may experience angina or heart failure, claudication, severe gastrointestinal dysfunction, or marked central nervous system symptoms even when the hemoglobin level is 10 g/dL.

Historical events related to the onset of the anemia should be carefully explored, and some indication of duration should be sought by determining the onset of symptoms, past hemoglobin measurements, or transfusion history. Long-standing anemia raises the possibility of a hereditary disorder. In this case, additional information concerning family members is desirable. Race is also a potential clue in detecting the presence of an abnormal hemoglobin or an enzyme-deficiency state. Occupational and environmental histories, including medications and exposure to toxic chemicals, may provide an explanation.

A detailed review of systems is also essential. The presence of general symptoms of illness such as weight loss, fever, and night sweats suggests an underlying inflammatory state; a careful review of the gastrointestinal system may reveal a history of dysfunction or possible blood loss. The patient's nutritional habits, especially the level and frequency of alcohol intake, merit special attention. Alcohol is an important etiologic factor in a number of anemias, particularly when it interferes with adequate diet. Some complaints suggest a specific diagnosis. Patients with marked iron deficiency may crave ice or, less frequently, complain of a sore mouth or difficulty in swallowing. Patients with sickle cell anemia complain of episodic bone or abdominal pain.

On physical examination, the most direct indication of anemia is pallor, since there is a general relationship between skin and mucous membrane color and hemoglobin level. Temperature can have a marked effect on skin color because of vasoconstriction or vasodilatation, as can variations in melanin pigmentation. Age and disease can also be factors. Patients with myxedema or nephrotic syndrome are pale because the blood in their skin is partially obscured by subcutaneous fluid. The best areas for evaluation of hemoglobin concentration are the conjuctiva, mucous membranes, nail beds, and palmar creases of the hand.

Bleeding can cause both hypovolemia and anemia. When bleeding is acute, hypovolemia is the primary concern; when it is chronic, iron deficiency anemia may be the outstanding manifestation. With acute blood loss, initial compensation is provided by vasoconstriction in proportion to the degree of hypovolemia.When the volume loss approaches 30% of the total blood volume, compensation can no longer be achieved, and both postural hypotension and tachycardia with exercise are observed. An acute blood volume loss in excess of 35% to 40% is associated with decreased cardiac output and is manifested by the symptoms of hypovolemic shock, including anxiety, air hunger, sweating, mental confusion, hypotension, and tachycardia at rest.

When blood loss anemia develops gradually, an increase in the plasma volume keeps the total blood volume at near-normal levels, and oxygen delivery is maintained by compensating mechanisms such as increased cardiac output, a shift in the hemoglobin-oxygen dissociation curve, and redistribution of blood flow (see page 26). On physical examination of the severely anemic patient, the physician may detect a forceful apical impulse, hyperactive heart sounds, and a wide pulse pressure caused by the increase in ventricular stroke volume. Murmurs are often produced by the greater force of contraction and resulting blood turbulence. Generally, these are midsystolic or holosystolic flowing murmurs, best heard at the apex and along the sternal border with radiation to the neck.

Clues to the cause of the anemia may be found on physical examination. Patients with vitamin B_{12} deficiency may have stomatitis, glossitis, and papillary atrophy of the tongue. Progressive neurologic findings in this disease, including ataxia, loss of vibration and position sense, and changes in deep tendon reflexes, are indicative of progressive nerve fiber demyelination. Patients with increased blood breakdown caused by either hemolytic anemia or ineffective erythropoiesis may be slightly jaundiced because of increased bilirubin pigment. However, severe jaundice suggests significant liver disease. Splenomegaly may cause or be caused by increased red cell destruction in that organ. Lymph node enlargement is common in patients with malignancies of the lymphatic system and may be associated with an immune hemolytic state. A hemorrhagic state can be manifested by the presence of severe epistaxis, gum bleeding, petechiae, ecchymoses, or hematomas, or by a positive test for occult blood in the feces.

Laboratory Diagnosis

Although the presence of anemia may sometimes be suspected from the history and physical examination, the routine blood count is a far more definitive measure of the status of the erythron. Even if anemia is suspected, a full laboratory evaluation is essential to validate the diagnosis, determine its severity, and define its nature. The hematology laboratory offers a set of routine or standard procedures relevant to the diagnosis of anemia (Table 4).The most important of these are the complete blood count using an automated counter, blood smear morphology, reticulocyte production index, and evaluation of iron supply. Usually the complete blood count and smear are routine, whereas the reticulocyte index and iron studies are ordered if anemia is demonstrated and its cause is not obvious.

Table 4 • **Routine Laboratory Measurements in the Diagnosis of Anemia**

Complete blood count:
  Hemoglobin, hematocrit, red blood cell count, cell indices (mean cell volume, mean cell hemoglobin, mean cell hemoglobin concentration, red cell distribution width), white cell count and differential, and platelet count
Red cell morphology—smear description
Reticulocyte production index
Bilirubin and lactate dehydrogenase levels
Serum iron, total iron binding capacity, serum ferritin level
Bone marrow aspiration or biopsy, including marrow iron stain

Routine Blood Cell Measurements

The complete blood count (CBC) is best performed with an automated counter capable of simultaneous measurements of the hemoglobin, red cell number, red cell indices, platelet count, white blood cell count, and a three- or five-part white blood cell differential. Excellent instruments are now manufactured by Technicon, TOA, and Coulter Electronics. These counters use an electrical field or a highly focused light source to detect the presence and characteristics of individual cells in solution. The instrument automatically pipettes, dilutes, and then passes the dilute solution of blood cells through an aperture at a fixed rate to count individual cells. Since red blood cells are relatively poor conductors of electricity, they produce a momentary drop in conductance, the magnitude of which is a measure of their relative size. Therefore, the electrical impedance technique makes it possible to both rapidly and accurately count a large number of cells and simultaneously characterize their volume. A sample set of results from a Model S Coulter counter is shown in Figure 23. Counters that use laser optics provide the same measurements based on light scatter characteristics.

Since a large number of cells are counted very quickly, the measurement error (coefficient of variation) of automated counters is less than 2%. It is also possible to standardize the counter with a stable solution of fixed red cells and a known hemoglobin standard. Thus, the principal measurements, hemoglobin level, red cell count, and mean cell volume, can be standardized independently for greater accuracy.

The hemoglobin and hematocrit are the most commonly used measurements in the clinical detection of anemia. Since they are in

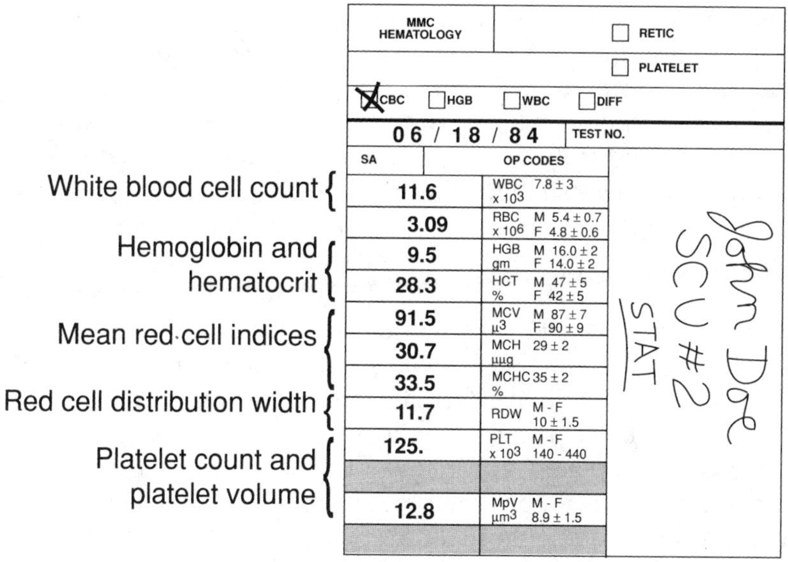

MMC HEMATOLOGY

☐ RETIC
☐ PLATELET
☒ CBC ☐ HGB ☐ WBC ☐ DIFF

06 / 18 / 84 TEST NO.

SA — OP CODES

Label	Result	Op code	Normal range
White blood cell count	11.6	WBC x 10^3	7.8 ± 3
	3.09	RBC x 10^6	M 5.4 ± 0.7, F 4.8 ± 0.6
Hemoglobin and hematocrit	9.5	HGB gm	M 16.0 ± 2, F 14.0 ± 2
	28.3	HCT %	M 47 ± 5, F 42 ± 5
Mean red-cell indices	91.5	MCV μ^3	M 87 ± 7, F 90 ± 9
	30.7	MCH μμg	29 ± 2
	33.5	MCHC %	35 ± 2
Red cell distribution width	11.7	RDW	M - F 10 ± 1.5
Platelet count and platelet volume	125.	PLT x 10^3	M - F 140 - 440
	12.8	MpV μm^3	M - F 8.9 ± 1.5

John Doe
SCU #2
STAT

Figure 23. Results of blood cell counts measured on a Model S Coulter counter. Primary measurements include the white blood cell count, the red blood cell count, the hemoglobin, the mean cell volume (MCV), the platelet count, and the mean platelet volume. In addition, from these values, the counter calculates the hematocrit, mean cell hemoglobin (MCH), mean cell hemoglobin concentration (MCHC), and red cell distribution width (RDW). A range of normal values with 95% confidence limits is provided on the printout. These values can vary slightly for individual laboratories, depending on the standardization of the counter and the nature of the patient population.

fair agreement (hemoglobin ×3 = hematocrit), it is largely a matter of physician preference as to which value is used. In terms of relative accuracy, the hemoglobin level is directly measured by automated counters and offers the most precise measurement of the oxygen carrying capacity of blood. The hematocrit is not measured but is a calculated value from counter measurements of red cell number and mean cell volume. This can make the hematocrit somewhat less accurate. The red cell count can be affected by major increases in the white blood cell count to levels in excess of 100 $\times 10^3$ per μL. Furthermore, the mean cell volume (MCV) of the red cell population is influenced by cell agglutination and osmotic changes secondary to hyperglycemia and hypernatremia. In these situations, a sudden swelling of the red cell count can falsely elevate a hematrocrit calculated as a product of the red cell count and MCV.

To identify an abnormally low hemoglobin or hematocrit value, it is necessary to establish a mean normal value and the limits of normality. This depends not only on the apparent error of the measurement, but also on the expected variation between and within individuals. For example, although the hemoglobin determination on a single blood sample can be measured with the precision of ±0.1 g/dL, there is a much greater variation among a population of normal individuals. The hemoglobin values observed in a large population of adult males follow a symmetrical (Gaussian) distribution with a mean value of 15 g/dL of whole blood and a standard deviation of ±1 g. This means that two thirds of normal adult males have a hemoglobin value between 14 and 16 g/dL, whereas 95% fall between 13 and 17 g/dL. These outer limits, representing two standard deviations (2 SD) from the mean, are often referred to as the "normal range."

The fact that a patient's hemoglobin level falls just outside of this range does not necessarily mean that it is abnormal. Because of the overlap between abnormal and normal individuals, this interpretation can never be more than a probability (Fig. 24). The lower the hemoglobin value, the more likely it is that the value represents true anemia. At the same time, because of the overlap, it is possible to be anemic but have a hemoglobin value that falls, yet remains within the 2 SD range. To further complicate the problem, if the prevalence of anemia in a population changes, the probability and, therefore, the sensitivity of the measurement also change.

Additional factors affect any given hemoglobin measurement. The conditions under which the blood is drawn can alter the value obtained. For example, if the patient is in an upright position, the hemoglobin concentration is about 0.7 g/dL higher than if he or she is supine. Undue anxiety or excessive pain associated with venipuncture can cause catecholamine discharge and vasoconstriction, resulting in immediate reduction in plasma volume and an increase in the hemoglobin concentration by as much as 1 g/dL. Likewise, changes in fluid balance may affect plasma volume and, thereby, hemoglobin concentration.

More important is the normal variation caused by age and sex (Table 5). Various reasons have been proposed for these differences. It is suggested that, in childhood, higher inorganic phosphate levels result in increased red blood cell 2, 3-DPG, and therefore in lower hemoglobin values.[40] Postpubertal young men have higher hemoglobin values secondary to the stimulatory effect of androgens on erythroid precursors. Placental hormones may be responsible for a hemoglobin decline of about 1 g/dL during the second and third trimesters of pregnancy, and wide swings in extra-

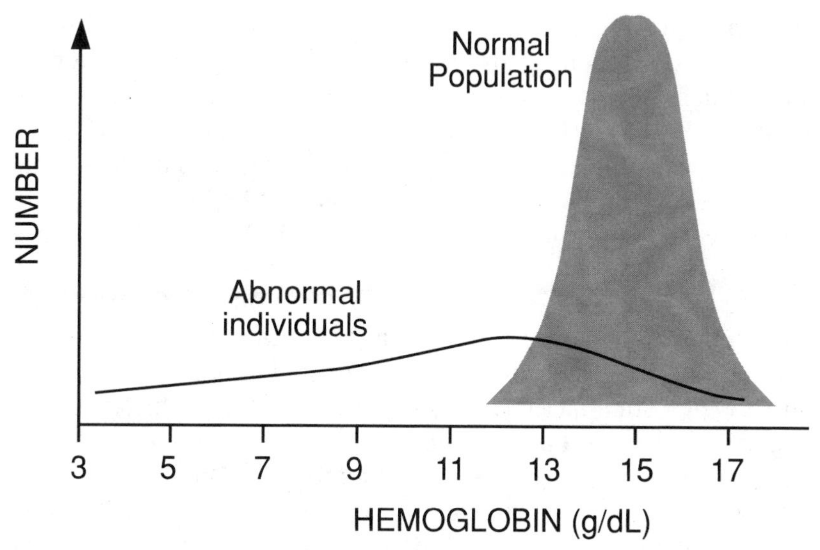

Figure 24. Distribution of normal and abnormal hemoglobin values. The hemoglobin values observed in a large population of normal individuals follow a symmetrical (Gaussian) distribution. When this pattern is present, it is common practice to state the mean normal value (15 g/dL) and the range of normal as either one or two standard deviations around the mean. At two standard deviations (15 ± 2 g/dL), 95% of normal individuals are included. Thus, the fact that a male patient's hemoglobin is slightly below 13 g/dL does not necessarily mean that it is abnormal—there is still a small chance that it is normal. The lower the value, the more likely it represents true anemia. Patients with abnormal hemoglobin values may not be detected because their hemoglobin values fall within the normal range.

Table 5 • **Mean Normal Values**

Age	**Hemoglobin** (g/dL)	**Hematocrit** (%)
Birth	17.0	52
1 to 3 months	14.0	42
3 months to 5 years	12.0	36
6 to 10 years	12.0	37
11 to 15 years	13.0	40
Adult man	15.0	46
Menstruating woman	13.5	41
Pregnancy (2nd and 3rd trimesters)	12.0	37

cellular fluid volume can make the fall even more pronounced. Racial differences may also be associated with different hemoglobin values; normal values for blacks in the United States are approximately 0.5 g/dL below those of whites.[41] Finally, a number of environmental factors can play a significant role, particularly those affecting oxygen supply. Altitude has a predictable effect with an increase of 1 g/dL of hemoglobin for each 3% to 4% decrease in arterial oxygen saturation. Smoking produces sufficient carbon monoxide to decrease hemoglobin oxygen saturation and increase hemoglobin concentration. A heavy smoker may increase his or her hemoglobin level by 0.5 to 1 g/dL. The clinician, therefore, must weigh such factors for each patient and differentiate between a physiologic (true) anemia and a laboratory anemia.

The indices of red cell size and hemoglobin content are very important in the differential diagnosis of anemia (Table 6). The MCV is the most valuable. In the nonanemic subject, the normal MCV is 90 ± 5 fL. Values greater than 100 fL usually signify a nuclear maturation defect (see page 96). Values less than 85 fL suggest impaired hemoglobin synthesis, and values of less than 80 fL are diagnostic of a hemoglobin synthesis defect. When performed on modern automated counting equipment, the MCV is highly reproducible, yet er-

Table 6 • **Red Cell Indices and Anemia Diagnosis**

Normocytic/Normochromic
- MCV—90 ± 5 fL; MCH—32 ± 2 pg
  - Hypoproliferative anemias: marrow damage, renal disease, inflammation, early iron deficiency
  - Hemorrhagic/hemolytic anemias

Microcytic/Hypochromic
- MCV <85 fL; MCH <30 pg
  - Iron deficiency anemia
  - Thalassemias
  - Hereditary sideroblastic anemia

Macrocyctic/Normochromic
- MCV >100 fL; MCH >30 pg
  - Vitamin B_{12} or folic acid deficiency
  - Intrinsic marrow disease
  - Chemotherapy
  - Liver disease

rors may occur. A falsely high MCV can be caused by extremely high numbers of white cells, osmotic effects caused by alterations in blood sugar or sodium as described under the hematocrit discussion, and clumping of red cells when cold agglutinins are present.

The mean cell hemoglobin (normal MCH = 32 ± 2 pg) is usually closely tied to the MCV and increases or decreases with the MCV. Exceptions are rare and seen only when microcytosis is produced by red cell fragmentation or osmotic changes. The mean cell hemoglobin concentration (normal MCHC = 33 ± 2%) is insensitive to changes in red cell hemoglobin content and of little value in separating impaired hemoglobin synthesis associated with iron deficiency from that with other causes. An increase in the MCHC to levels greater than 35% can be caused by the presence of large numbers of spherocytes, cells that are microcytic because of a loss of both intracellular water and membrane.

Automated counters also calculate an index of cell size variation, termed the *red cell distribution width* (RDW). Two different calculated values are provided. The RDW-CV is measured as a ratio of the width of the distribution curve at one standard deviation divided by the MCV (Fig. 25). The RDW-SD is a direct measurement of the distribution width at the 20% frequency level. Normally, the size distribu-

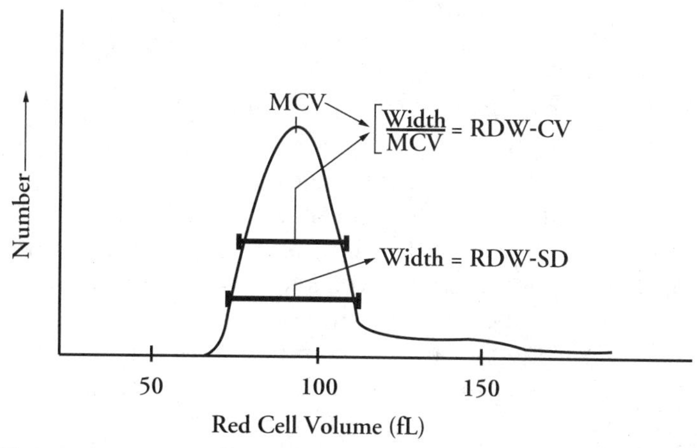

Figure 25. Red cell distribution width (RDW) measurements. As part of the CBC, electronic cell counters provide two measurements of the relative distribution of red cell size, the RDW–CV and the RDW–SD. The RDW–CV is calculated as the ratio of the red blood cell histogram width at one standard deviation divided by the MCV. The RDW–SD is simply the histogram width at the 20% frequency level.

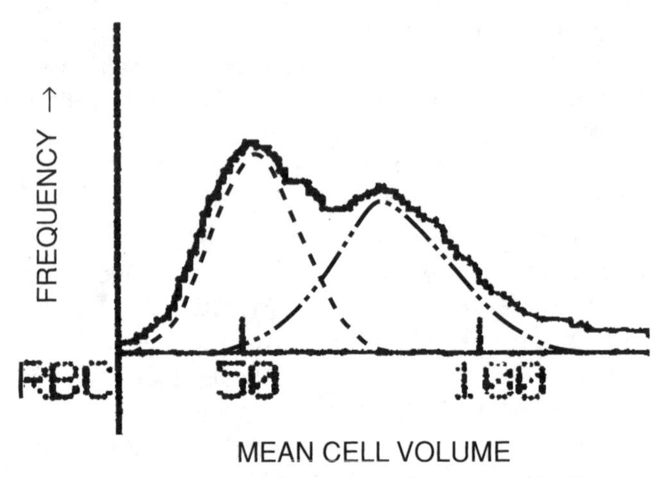

Figure 26. Size distribution curve for distinct populations of red cells. Inspection of the size distribution curve provided by the automated counter may reveal the presence of distinct populations of red cells of different cell volume. This example shows a curve for a patient with a population of extremely small cells (*dashed line*) mixed with a population of near normal sized cells (*broken line*).

tion curve for red cells is quite symmetrical, with an RDW-CV value of 10 ± 1.5% and an RDW-SD of 42±5 fL. A high RDW—that is, a greater variation in cell size—is caused by either the appearance of macrocytic or microcytic cells or the admixture of two distinct populations of cells.[42] The nature of the variation in cell volume can be determined by direct inspection of the size distribution plot (Fig. 26).

Red Cell Morphology

Although an automated counter provides a highly accurate method for detecting average changes in red cell size and hemoglobin concentration, it is less sensitive to the presence of minor populations of abnormal cells. Additional clues to the nature of the anemia, such as changes in red cell shape and staining characteristics and the appearance of immature or abnormal cells, can be better identified through a careful examination of the blood smear.

The Wright's stain blood smear must be technically adequate, since even expert morphologists cannot interpret a poor smear. A small drop of blood from a fingerstick or fresh anticoagulated blood sample is placed on one end of a clean glass slide and, as shown in Figure 27, a second slide is then used to draw the blood the length of the slide with a smooth motion. The thickness of the smear can be varied by changing the angle of the applicator slide and the speed

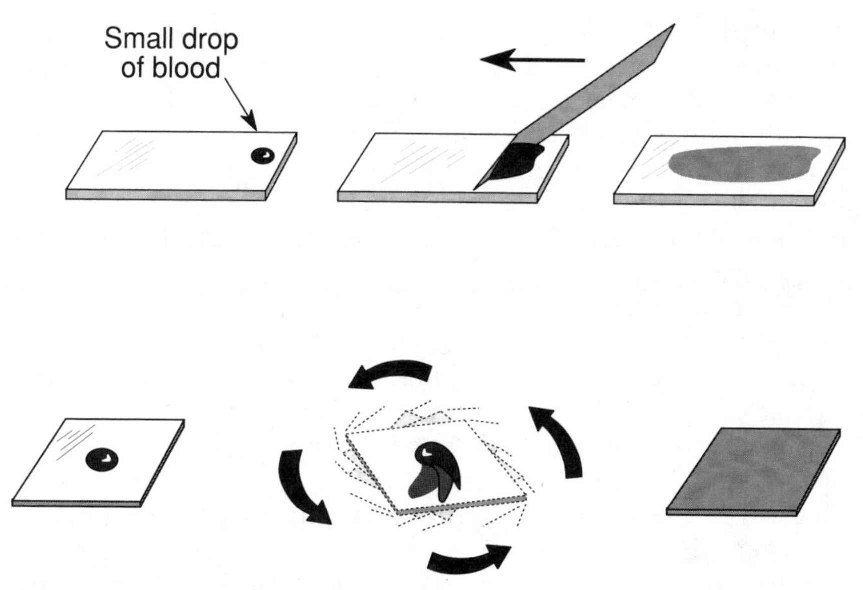

Figure 27. Preparation of the blood smear. Blood smears may be prepared using coverslips and a slide spinner or glass slides. In the latter case, a small drop of blood from a fresh anticoagulated blood sample or finger stick is placed at one end of the clean glass slide. Using a smooth motion, the edge of a second slide is then used to draw the blood along the length of the slide. The thickness of the smear varies according to the severity of the patient's anemia. It is also possible to vary the thickness by changing the angle of the applicator slide and the speed of the motion. After the blood is air dried the slide is stained with Wright's stain. The slide is flooded with stain for 2 to 3 minutes and then a buffer solution is added until a green sheen appears on the surface. After another 3 to 5 minutes, the slide is rinsed with tap water and air dried.

of the motion. Smears may also be made with coverslips, a technique that offers somewhat better red cell morphology. In this case, a tiny drop of blood is placed in the center of the coverslip, a second coverslip placed over the drop of blood, and the blood permitted to spread. The two coverslips are then rapidly pulled apart. After the blood is air dried, the glass slide is flooded with Wright's stain for 2 to 3 minutes. A buffer solution is then added until a green sheen appears on the surface. The slide is left to stand for 3 to 5 minutes, after which it is rinsed with tap water and air dried. Once stained, smears are stable and can be stored almost indefinitely.

When a blood smear is examined under the microscope, care must be taken to find that portion of the smear where red cell morphology is optimal. Generally, the best area is in the thinner portions of the smear, where red cells are nearly touching but not overlapping and most are biconcave (have pale centers). For glass slide smears, this is usually the area near the tail end of the smear. With coverslip smears, it is generally in the center. White cells, especially lymphocytes, are often drawn to the leading and outer edges of the smear, distorting the distribution of both elements. Despite this potential problem, the numerical relationship between red cells, white cells, and platelets is worth evaluating under low power magnification (Fig. 28).

Examination of the blood smear of the anemic patient permits certain general interpretations. A smear that shows little change in erythrocyte size or shape is commonly seen with hypoproliferative anemias. On the other hand, a smear with dramatic size and shape changes, marked macrocytosis, microcytosis, or poikilocytosis, is indicative of either ineffective erythropoiesis or fragmentation hemolysis. The presence of polychromatic macrocytic erythrocytes (shift cells) is an indicator of increased erythropoietin stimulation (see Fig. 34). However, when these immature red cell forms are

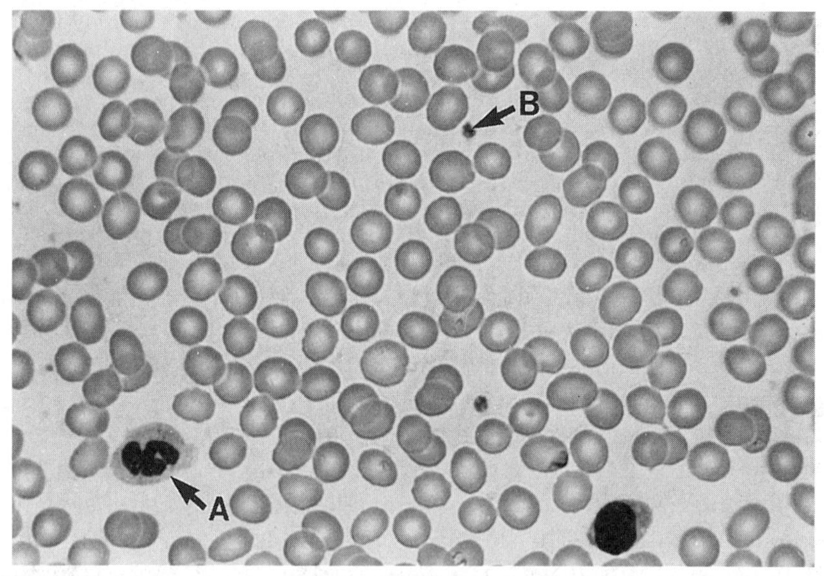

Figure 28. Blood smear morphology. The best area on blood smear for the examination of red cell morphology is the thinner portion, where red cells are nearly touching but not overlapping. Under low power magnification, it is possible to evaluate the relative frequency of white blood cells *(arrow A)* and platelets *(arrow B)* in relationship to the number of red blood cells. There should be approximately one white cell for every 500 to 1000 red cells and one platelet for every 5 to 10 red cells.

also associated with other immature blood cells types, it suggests stromal abnormalities in the marrow, such as occurs with fibrosis or tumor invasion (myelophthisic blood picture).

The blood smear complements the red cell indices and RDW (see Figs. 25 and 26). Overall size changes are best reflected by the MCV. However, as an anemia develops, the MCV changes slowly because of the time lag in turnover of circulating red cells. At the same time, the smear reveals minor populations of either smaller or larger red cells, which have yet to affect the MCV (Fig. 29). For example, in a patient who is just becoming iron deficient, examination of cells freshly released from the marrow may allow detection of the decrease in hemoglobin synthesis and reduction in cell volume before an overall decrease in the MCV is apparent. Likewise, an evolving megaloblastic anemia is associated with the appearance of occasional macrocytes on smear before there is a significant increase in the MCV.

Variations in cell size without a change in cell shape is termed *anisocytosis.* To interpret the meaning of anisocytosis, the most important consideration is whether there is a mixture of normal cells and small cells, of large cells and normal cells, or of all three vari-

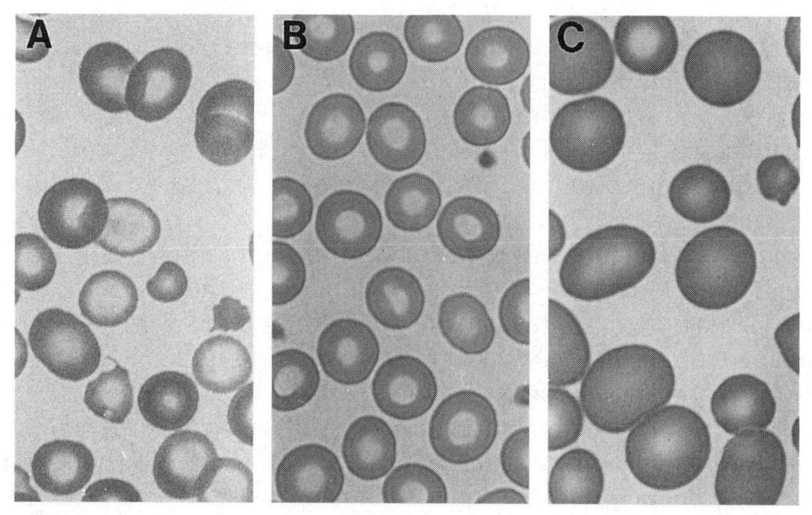

Figure 29. Red cell size changes as detected by morphology. When pronounced, microcytosis and hypochromia (*A*) and macrocytosis (*C*) are readily distinguishable from normocytic, normochromic red cells (*B*). The blood smear can be more sensitive than the MCV by revealing a minor population of either smaller or larger cells that have little effect on the MCV.

eties. The presence of small, hypochromic cells indicates a defect in hemoglobin synthesis. When macrocytes are present, one must further deduce whether the large cells are shift cells, true macrocytes, or the macrocytic target cells of liver disease (Fig. 30, see also Figs. 29 and 34). The presence of true macrocytes usually indicates a defect in nuclear maturation, even if small cells are also present. Since macrocytosis caused by a nuclear maturation defect is often accompanied by poikilocytosis, it is not unusual to see a variety of cell sizes with megaloblastic anemias.

Changes in red cell thickness are reflected by a change in the density of the eosinophilic stain and, often, an absence of central pallor (see Fig. 30). Some increase in density of staining is to be expected in larger cells, even though the concentration of hemoglobin is normal. Microcytes that stain more densely usually have both a decrease in red cell membrane and an increase in hemoglobin concentration. These cells are called *spherocytes.* At the opposite extreme are flattened, pale cells. When cells show increased pallor and a central pink deposit, they are referred to as *target cells.* Flattened or target cells have excessive cell membrane in comparison to their hemoglobin content, the result of impaired hemoglobin synthesis or an abnormal accumulation of cell membrane. Targeting is most conspicuous with certain hemoglobinopathies, especially hemoglobins C and S, but also occurs with hepatic disease and hyposplenism.

Cell fragmentation occurs with either ineffective erythropoiesis (where cell breakage may occur as new cells emerge from the marrow through the sinusoidal wall) or mechanical damage to circulating red blood cells caused by abnormalities of vascular endothelium (microangiopathic hemolytic anemia) or environment (thermal burn, mechanical heart values, and so on). Distorted cell shapes are also seen with structural or metabolic abnormalities that affect the maintenance of the cell membrane. In fact, some diseases can be diagnosed because of the presence of a characteristic morphologic abnormality. Sickle cell anemia is one such example. Hereditary elliptocytosis is another. Certain shape changes and the presence of intracellular inclusions provide information regarding cell production and destruction, especially the role of the spleen and reticuloendothelial system in clearing defective red cells. The presence of nuclear remnants (Howell-Jolly bodies) may indicate hypofunction or absence of the spleen (hyposplenism). Teardrop and nucleated red cells are often associated with myelofibrosis and extramedullary hematopoiesis. Further discussions of morphologic

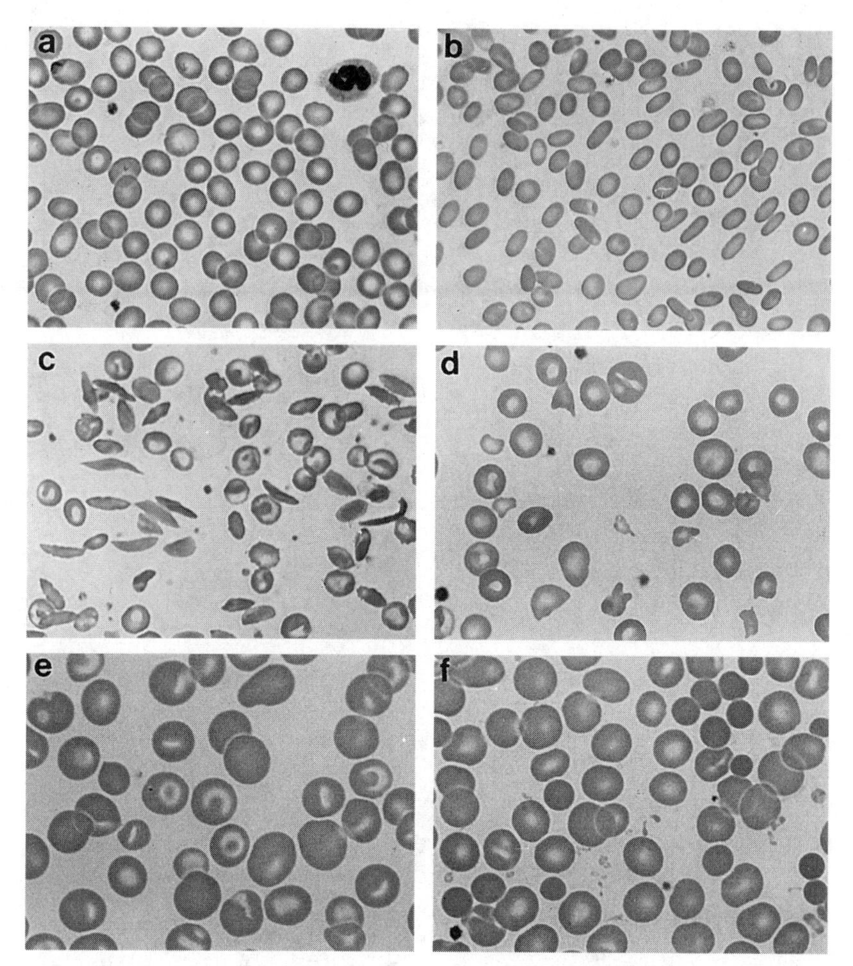

Figure 30. Abnormal red cell morphology. The blood smear can provide important morphologic clues to red cell disease. A number of important morphologic abnormalities are shown in this composite, including (*a*) normal smear for comparison; (*b*) elliptocytosis from a patient with hereditary elliptocytosis; (*c*) sickle cells from a patient with Hb S thalassemia; (*d*) red cell fragmentation from a patient with a heart valve prosthesis; (*e*) target cells and uniform macrocytosis from a patient with liver disease; (*f*) spherocytosis from a patient with hereditary spherocytosis; (*g*) teardrop cells and stippling in a patient with myelofibrosis; (*h*) acanthocytosis from a patient with liver disease; (*i*) siderocytes from a splenectomized patient with thalassemia; (*j*) a nucleated red blood cell from a patient with marrow damage; (*k*) a red cell containing a Howell-Jolly body (a residual nuclear fragment) from a splenectomized patient; and (*l*) a red cell containing a malarial parasite.

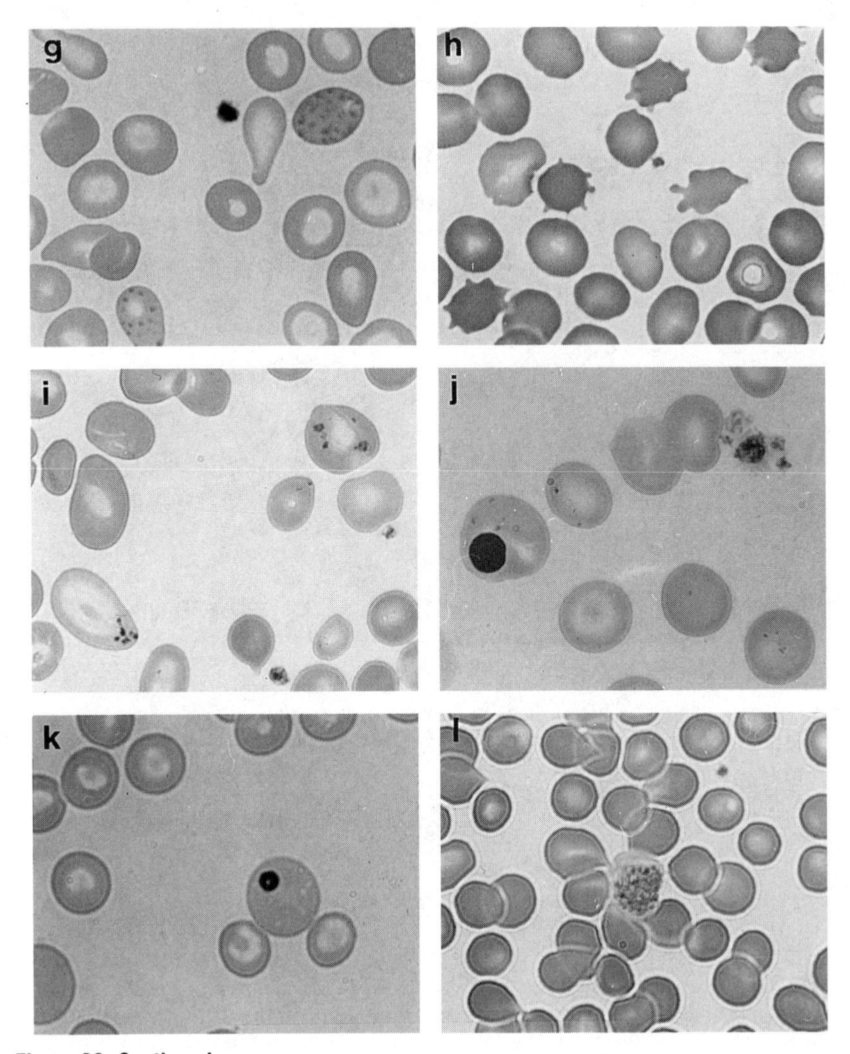

Figure 30. Continued

abnormalities in later sections deal with the diagnosis and management of specific disorders.

Marrow Examination

In an anemic patient with an abnormal blood smear, examination of the bone marrow provides important additional information. A marrow specimen can be easily obtained from the sternum or anterior iliac crest and, in infants, from the upper tibia, but the pre-

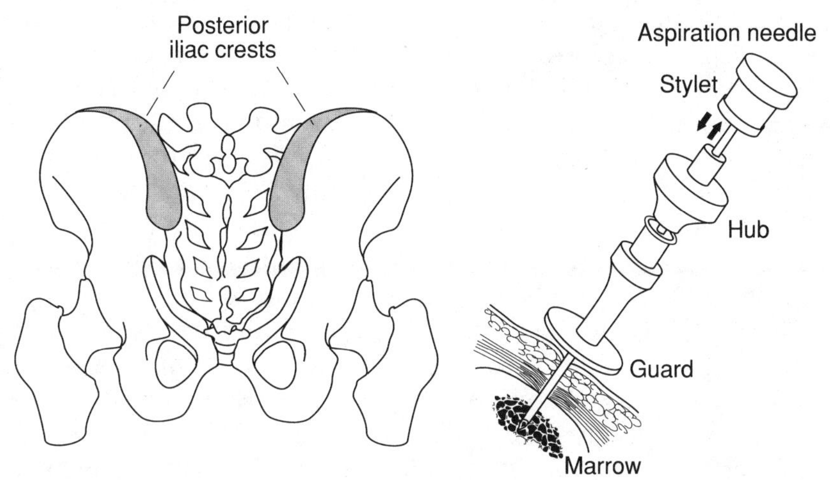

Figure 31. Bone marrow aspiration. The preferred site to obtain bone marrow in the adult is from the posterior iliac crest, which may be palpated as a bony ridge approximately 2 cm from the midline at the level of the sacrum (*shaded area*). At this location, the marrow cavity is several centimeters deep, making it possible to penetrate the cavity and aspirate its contents with little risk of soft tissue or organ damage. A standard aspiration needle is shown in the right-hand panel. With the stylet in place, the needle is pushed through the external bony plate using a twisting motion. A give in resistance and momentary discomfort as the needle penetrates the endosteum indicates that the needle has entered the marrow cavity. The stylet is then withdrawn and a 1- or 2-mL syringe is used to aspirate a small amount of marrow. To ensure a rich sample, only a few drops of marrow should be aspirated. Otherwise, the sample will be heavily contaminated with venous blood. The needle may be twisted in place to disrupt the marrow structure and dislodge small spicules of particulate marrow. If no specimen is obtained, a marrow biopsy may be required to detect abnormalities of marrow structure or marrow densely packed with tumor cells.

ferred location in adults is the posterior iliac crest (Fig. 31). For evaluating cell morphology, a smear of aspirated marrow particles is the most useful. Marrow particles can also be embedded in a clot and sectioned, providing an opportunity to examine marrow structure, especially the relationship of hematopoietic cells to marrow stroma.

The adequacy of the marrow specimen is critical. The best aspirate smears are made by harvesting several marrow particles from the specimen and spreading them on slides or coverslips followed by Wright's and Giemsa staining. With a good preparation, both the structure of the stromal particle and the morphology of individual cell types are apparent. Smears that lack particulate matter or show large numbers of distorted or fragmented cells are inadequate for interpretation.

Examination of a marrow particle smear begins with a low-power scan for overall cellularity and the relative frequency of megakaryocytes. The latter cells in their adult form are easily recognized since they are very large and have multiple nuclei. An estimate is then made under high-power magnification of the ratio of erythroid to granulocyte precursors (the E:G ratio). Normally, the E:G ratio is approximately 1:3. In situations where white cell production is at a basal level (circulating granulocyte count is normal), the E:G ratio can be used as an index of red cell production (page 38). Next, the sequence of red cell maturation is examined with attention to cell size, nuclear development, cytoplasmic characteristics, and hemoglobin content. From this assessment, it is possible to detect specific abnormalities in nuclear and cytoplasmic development. For example, stimulation by erythropoietin produces a macronormoblastic marrow characterized by foci of normoblasts in nearly synchronous maturation with moderate enlargement of the nuclei and somewhat immature cytoplasm. In contrast, megaloblastic erythropoiesis is associated with much larger nuclei showing a fine chromatin pattern and a discrepancy between nuclear maturation and hemoglobin synthesis. The latter has been referred to a nuclear-cytoplasmic dissociation, in which the nucleus looks less mature than it should for the amount of hemoglobin synthesized. Distortion and fragmentation of nuclei are also common, reflecting failure of red cell precursors to complete the maturation sequence. Impairment of hemoglobin synthesis results in increased numbers of normoblasts with small pyknotic nuclei and a small amount of transparent cytoplasm, so-called "tissue paper" normoblasts.

Failure to obtain cellular material by aspiration, provided the needle is in the marrow cavity, suggests either an empty marrow, increased reticulin with fibrosis, or dense packing by tumor cells. In these situations, or for the evaluation of marrow structure, a needle biopsy should be performed.[43] Total cellularity varies with age. The density of hematopoietic tissue in the central skeleton decreases from about 65% of the available marrow space in the young adult to approximately 30% in the elderly population. The remaining space is filled with fat cells. This change in cellularity can be best evaluated from a biopsy specimen. Biopsy is also the preferred technique for the detection of focal granulomatous lesions or collections of tumor cells. Unfortunately, the required decalcification of the specimen distorts the morphology of individual hematopoietic precursors, making their recognition more difficult.

The Prussian blue stain of marrow-aspirate smears or biopsy material for iron is very helpful in the differential diagnosis of anemia. It permits evaluation of both reticuloendothelial-cell iron stores and iron deposition in individual normoblasts.[44,45] In normal individuals, approximately 30% of normoblasts normally contain one to three small blue-staining ferritin granules. These cells are called *sideroblasts.* Marrow reticulocytes may also contain iron granules (*siderocytes*). A decrease in the number of sideroblasts suggests iron-deficient erythropoiesis, whereas an increase indicates an iron uptake in excess of the hemoglobin-synthesizing capacity of the developing red cell. Distinctly abnormal sideroblasts are observed in patients with globin or porphyrin synthesis defects (Fig. 32). Thalassemic patients, with a globin production defect, have two to ten blue granules, representing increased amounts of semicrystalline aggregates of ferritin in the cytoplasm of developing normoblasts (*ferritin sideroblasts*). Patients with abnormalities in mitochondrial function or porphyrin synthesis have ring sideroblasts, in which up to 30 blue granules, representing iron-encrusted mitochondria, form a ring around the nucleus (*mitochondrial sideroblasts*).

Measurements of Red Cell Turnover

There are four clinically useful laboratory measurements of red cell turnover: the reticulocyte count, the marrow E:G ratio, the serum

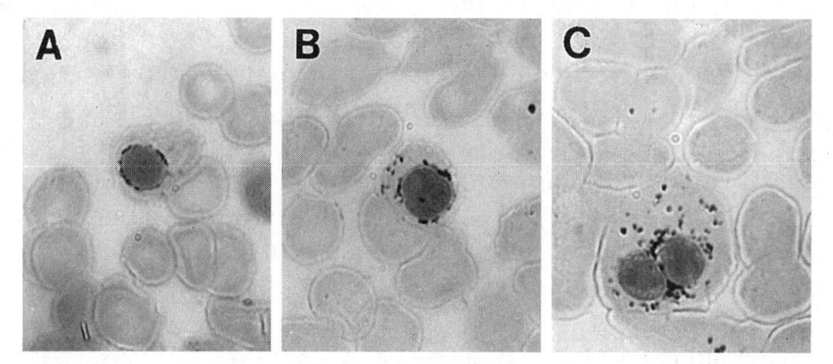

Figure 32. Abnormal and ring sideroblasts. In the normal subject, approximately 30% of normoblasts contain one to three small blue-staining ferritin granules in the cytoplasm. Abnormal sideroblasts are observed in patients with globin or porphyrin synthesis defects. (*A*) The typical ring sideroblast from the patient with a hereditary or acquired ring sideroblastic anemia. (*B*) Perinuclear and cytoplasmic iron deposits as observed with globin production defects. (*C*) Chronic alcoholism with both a megaloblastic anemia (the nuclear abnormality) and a sideroblastic defect.

bilirubin, and the serum lactate dehydrogenase (LDH) level. Of these, the reticulocyte count is the most practical for the clinical evaluation of red cell turnover.

The reticulocyte is an immature red cell containing ribosomal RNA, which can be precipitated by supravital dyes such as new methylene blue.[46] A fresh sample of blood is mixed with a few drops of new methylene blue solution and, after a 10- to 15-minute incubation, coverslip smears are prepared for counterstaining with Wright's stain. If possible, reticulocyte smears should be prepared using a blood smear centrifuge to guarantee a uniform distribution of cells. The reticulocyte count is then performed by counting 1000 red cells on each of two coverslips and noting the percent of cells containing more than two blue-staining particles (clumps of precipitated RNA).

Normal individuals have a reticulocyte count of about 1%. Since this is a relatively small number of cells, the standard error of the measurement is quite large. The normal range is usually given as 0.4% to 1.7%, largely reflecting the counting error. At higher reticulocyte counts, the accuracy of the method improves significantly. Since the principal objective of the reticulocyte count is to identify major increases in red cell production of three to five times normal, the accuracy of the method is adequate for clinical diagnosis.

To use the reticulocyte count as an index of red cell production, two corrections need to be made. The first is the conversion of the percent count to an absolute number. This is accomplished simply by multiplying the reticulocyte percentage by the red cell count, to obtain the absolute number of reticulocytes per microliter (μL) of blood. Normally, this is approximately 50,000 reticulocytes/μL. Alternately, the absolute reticulocyte percentage can be calculated by multiplying it by the patient's hemoglobin (hematocrit) divided by the normal hemoglobin (hematocrit). For example, an individual with a 3% reticulocyte count and a hemoglobin of 7.5 g/dL (hematrocit 23%) has an absolute reticulocyte percent of 1.5, as shown by the following calculation:

$$\text{Absolute reticulocyte count} = 3\% \times 7.5/15 \text{ (or } 23/45) = 1.5\%$$

The second correction is related to the maturation of the reticulocyte and its behavior with stimulation by erythropoietin. After the developing normoblast loses its nucleus to form a marrow reticulocyte, it takes 4 days for the ribosomal RNA to disappear from the developing cell. Usually, 3 of these days are spent in the marrow and only 1 day in circulation.[46] In the normal individual, the percent of reticulocytes found in circulation reflects the output of these "1-day" reticulocytes. However, with increased erythropoietin stimulation, the marrow maturation time becomes shorter and the blood maturation time for reticulocytes becomes longer. They now circulate for 2 or even 3 days before their blue-staining quality is lost. To obtain a true measure of red cell production, it is necessary to recognize that marrow reticulocytes have *shifted* out of the marrow to correct for their prolonged time in circulation.[32]

As a general rule, the correction for *reticulocyte shift* can be based on the severity of the anemia (Fig. 33). The prolongation of the maturation time is a function of the level of erythropoietin stimulation as determined by the severity of the anemia or hypoxia, and a correction factor can be derived according to the hemoglobin level. This factor is then used to carry out the second correction of the reticulocyte count. For example, for a reticulocyte

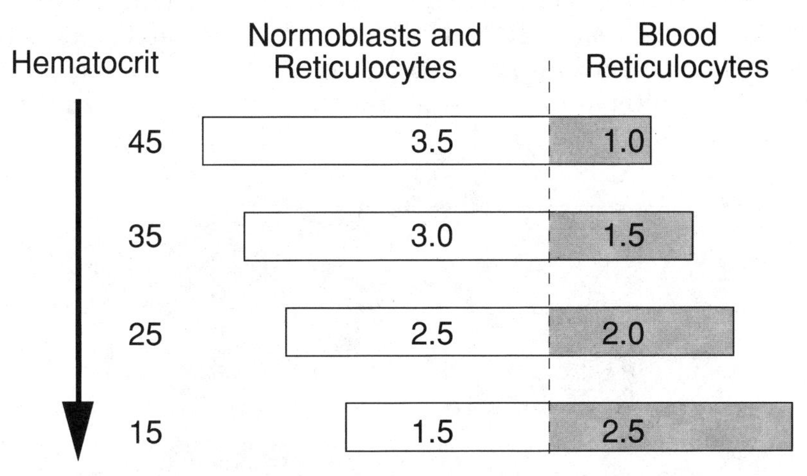

Figure 33. Correlation of hematocrit with the marrow and blood reticulocyte maturation times. Ordinarily, erythropoietin increases in proportion to the degree of anemia present. With increasing erythropoietin stimulation, the maturation time of the erythroid marrow normoblasts and marrow reticulocytes progressively shortens from a normal of 3.5 days to as little as 1.5 to 1.0 day(s). Much of this shortening is secondary to a shift of marrow reticulocytes into circulation. This results in a prolongation of the maturation time of circulating blood reticulocytes from a normal of 1 day to as much as 2.5 to 3.0 days with severe anemia. This needs to be taken into account when calculating the reticulocyte production index. The maturation time values shown for the blood reticulocytes can be used as a correction factor in this calculation (see text).

count of 9% in a patient with a hemoglobin of 7.5 g/dL (hematocrit of 23%):

Reticulocyte production index = 9% × 7.5/15 (hemoglobin correction) × 1/2 (maturation time correction) = 2.5

The use of the second correction factor assumes a predictable increase of erythropoietin in response to the anemia. As the hemoglobin decreases from 15 to 5 g/dL, the reticulocyte maturation time in circulation increases from 1 to as long as 3 days. It will be less than expected if there is a defect in erythropoietin stimulation, as seen in patients with renal disease or an inflammatory state. It will be more than expected if the patient is hypoxic for other reasons. The appearance of marrow reticulocytes in circulation can be validated by inspection of the Wright's-stained blood smear. Marrow or *shift* reticulocytes appear as polychromatic macrocytes; that is, they are larger and paler than normal and have a light blue staining quality (Fig. 34).[48]

The E:G ratio is an important companion to the reticulocyte index, providing a measure of the level of precursor proliferation. In the anemic patient with a reticulocyte index greater than three times normal, the E:G ratio should be greater than 1:1. This is indicative of

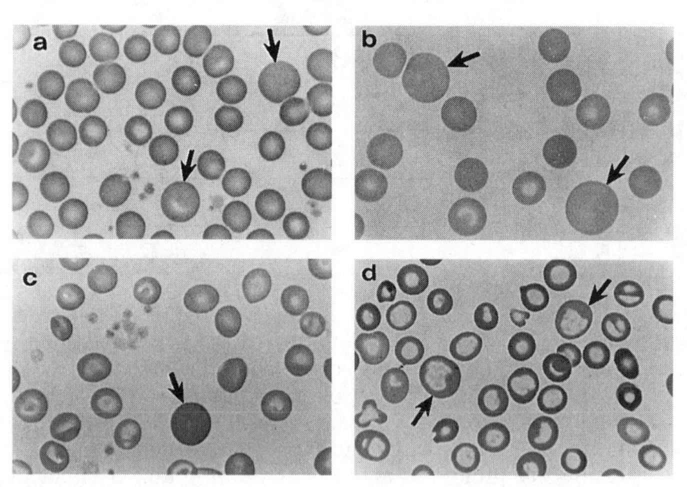

Figure 34. Shift cells. When a marrow reticulocyte appears in circulation, it can be recognized as a macrocytic cell, which is somewhat paler than normal, lacks a distinct area of central pallor, and, most importantly, has a light blue staining quality. Several examples are shown (marrow reticulocytes marked by arrows).

an *effective* proliferative response of the stimulated marrow. If an anemic patient has a reticulocyte index of less than 2 and an E:G ratio of greater than 1:1, red cell precursor proliferation is not accompanied by normal maturation. This is *ineffective erythropoiesis.*

The serum bilirubin and LDH levels provide semiquantitative measurements of cell production and destruction. The bilirubin concentration in blood, particularly the indirect fraction (*unconjugated bilirubin*), correlates with the rate of red cell breakdown. The normal total bilirubin level is between 0.4 and 1 mg/dL, of which 70% to 80% is unconjugated. Lower than normal values are found in patients with hypoproliferative anemias, particularly with iron deficiency or inflammation. Normal or slightly elevated values of 0.8 to 2 mg/dL are seen in patients with ineffective erythropoiesis and hemolytic anemia; the elevation is largely accounted for by an increase in the unconjugated fraction. The capacity of the liver to excrete conjugated bilirubin is so great that conjugated bilirubin is not elevated in patients with increased red cell turnover. A major increase in the conjugated bilirubin level indicates abnormal liver function.

The serum LDH level is even less specific, increasing with damage to a variety of tissues, including the liver and heart. However, increased serum levels are seen in patients with hemolytic anemia, especially those with intravascular hemolysis in whom the enzyme is cleared more slowly than hemoglobin. Serum LDH concentrations from 800 to 2000 IU/L (normal is 300 to 600 IU/L) are seen in patients with compensated hemolytic anemias. Elevations of several thousand units are seen with some examples of ineffective erythropoiesis, such as the megaloblastic anemias. In addition, there is a tendency to a reversal of the LDH isoenzyme pattern (LDH 1 exceeds LDH 2). Thus, the isoenzyme pattern can be used to indicate whether an LDH increase is secondary to hemolysis or to liver disease.

Tests of Iron Status

Iron supply plays a principal role in determining erythroid marrow proliferation and maturation. Since iron-deficient erythropoiesis is the most frequent kind of anemia, its evaluation is an essential component of the workup of the anemic patient. Measurements of plasma iron supply (plasma iron and total iron-binding capacity), iron stores (serum ferritin and marrow reticuloendothelial cell iron), and red cell iron supply (red cell protoporhyrin) contribute in specific ways to the evaluation of iron status.

The plasma iron and total iron-binding capacity with a calculation of the percent of transferrin saturated by iron are most important since they reflect the actual level of tissue iron supply. Determining the plasma iron (PI) involves a specific color reagent to quantitate iron after it is freed from transferrin (TF). The unsaturated or total iron-binding capacity (TIBC) is determined either colorimetrically or by isotopic measurement after sufficient iron is added to the specimen to saturate TF-binding sites. Normally, the PI fluctuates from an average value of 110 μg/dL plasma in the morning to 70 μg/dL in the evening (the normal range is 50 to 150 μg/dL plasma). This diurnal variation is unaffected by meals except in iron-deficient subjects. The TIBC (or TF level) remains relatively constant in the normal individual at 330±30 μg/dL but is altered in certain disease states (Fig. 35). An increase to greater than 370 μg/dL is a diagnostic of iron deficiency unless the individual is taking estrogens.

The most useful expression of plasma iron supply is the percent transferrin saturation (PI/TIBC).[45] This is normally 33±12%. When the transferrin saturation is less than 16%, hemoglobin synthesis can no longer be sustained at a basal level and the red cells will, in time, show the changes of iron deficiency. The fall in percent saturation may be caused by either an absolute deficiency of iron or an inflammatory state. However, the patterns of PI and TIBC are usually quite different in these two conditions. Patients with absolute iron deficiency tend to have a higher than normal TIBC (that is, transferrin level), while inflammation is associated with a decrease in the TIBC. A high percent saturation (greater than 60%) is occasionally seen as a normal fluctuation, but persistence of this deviation usually signals hepatic disease, abnormal erythropoiesis, or parenchymal iron overload.

Body iron stores may be evaluated by measurements of the serum ferritin level or by microscopic examination of marrow aspirate particles for hemosiderin. The serum ferritin level is measured by either radioimmunoassay or an enzyme-linked immunoassay.[49] The value of determining the serum ferritin level is that the amount of ferritin in circulation is usually proportionate to the amount of storage iron (ferritin and hemosiderin in body tissues). One μg/L of serum ferritin is equivalent to approximately 120 μg of storage iron per kilogram of body weight. This is true whether the iron is located in the reticuloendothelial cells or liver parenchymal cells. Serum ferritin levels vary with the sex and age of the individual in parallel to the changes in iron stores (Fig. 36). Normal

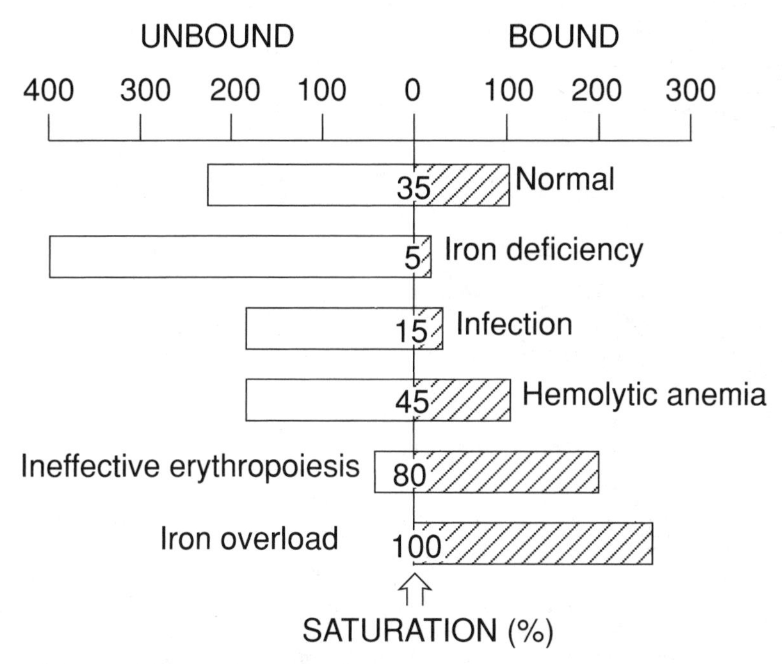

Figure 35. Transferrin and its iron. The serum iron in a normal individual ranges from 50 to 150 μg/dL of plasma, with 100 μg/dL being an average normal value. This represents only about one third of the binding capacity of transferrin, so that the saturation is about 30%. Both the serum iron and the total iron binding capacity (TIBC) levels are altered by disease states. Pregnancy is associated with a modest reduction in the serum iron and an increase in the TIBC. Iron deficiency shows an even more striking change with a major reduction in the serum iron and a dramatic increase in the TIBC, giving a percent saturation below 10%. In contrast, infection and inflammation reduce both the serum iron and the TIBC. The pattern observed with most hemolytic anemias is quite different from that observed with marked ineffective erythropoiesis. In the hemolytic anemias, the plasma iron is relatively normal, whereas with ineffective erythropoiesis there may be an increase in the serum iron level with near total saturation of the TIBC. A similar pattern is seen with hemochromatosis. Severe starvation, nephrotic syndrome, and chronic inflammatory disease can result in marked hypoproteinemia with pronounced reductions in transferrin protein.

levels for adult menstruating women range from 12 to 100 μg/L (mean 30 μg/L); adult males have values between 50 and 200 μg/L (mean 100 μgL) of plasma. A level below 12 μg/L indicates iron deficiency. If an anemic patient has a low transferrin saturation and an elevated ferritin level, inflammation is likely. The usual relationship between the serum ferritin level and body iron stores is disturbed by hypermetabolism, inflammation, damage to ferritin-rich tissues such as the liver, and the production of apoferritin by

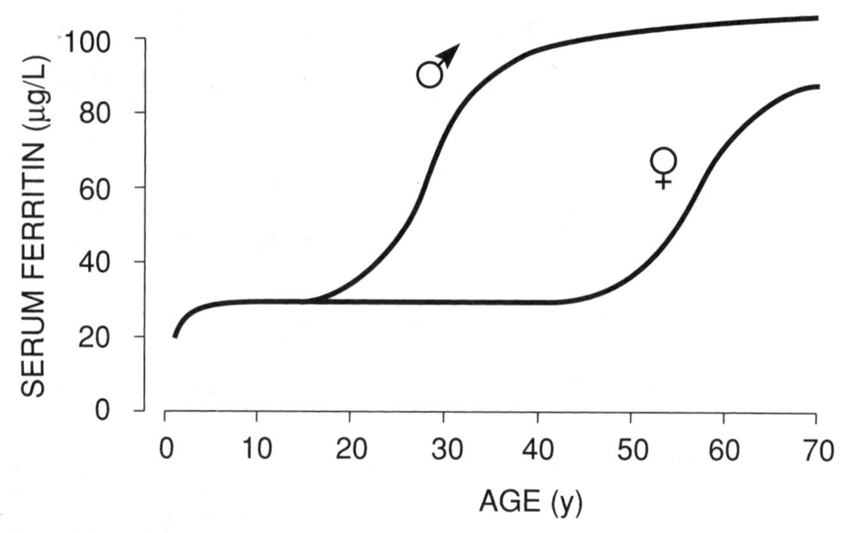

Figure 36. Serum ferritin levels according to sex and age. The serum ferritin reflects the marrow stores of the individual. However, the pattern of the serum ferritin level depends on the sex and the age of the individual. Before age 20, the serum ferritin level is usually between 20 and 40 µg/L. During the adult years, the values for men rise progressively during the early adult years to reach levels about 100 µgL. In contrast, values in most women remain between 20 and 40 µg/L throughout the childbearing years and increase only after menopause.

certain tumors.[49] All of these result in inappropriately high serum ferritin levels.

Marrow hemosiderin can be examined in unstained smears of marrow particles. Hemosiderin appears as yellow-brown refractile granules within the unstained reticuloendothelial cells. Preferably, aspirate smears, clot sections, or biopsy specimens can be stained with Prussian blue. It is essential that marrow particles be present and that the preparation be fixed in absolute methanol before staining with potassium ferrocyanide to prevent iron removal during the staining procedure.

Prussian blue is an extremely sensitive stain. With a good preparation, some iron will be visible when body iron stores are as low as 100 mg. The amount of reticuloendothelial-cell iron in a specimen is usually graded as 0 = absent, 1+ = small amounts (average woman), 2+ = moderate amounts (average man), 3+ = increased iron, and 4+ = massive iron deposits. The latter can be seen as a blue color on gross inspection of the slide. Additional information is provided by the size of individual particles within the reticuloendothelial cells. The presence of many small hemosiderin particles suggests a rapid iron turnover, as seen with a hemolytic anemia. The presence of large particles suggests a slow iron turnover and is characteristic of an inflammatory state.

The level of zinc erythrocyte protoporphyrin (EP) reflects the amount of protoporphyrin in excess of that used for hemoglobin synthesis.[50] It is easily measured fluorometrically. Because of difficulties in standardization, levels must be expressed according to values established by the individual laboratory. However, normal EP levels usually average about 35 μg per dL of red blood cells. Levels greater than 70 μg/dL are seen with an inadequate iron supply to the marrow (iron-deficient erythropoiesis), caused by body iron deficiency or inflammation. Increased EP levels also occur with lead poisoning and certain rare disorders of porphyrin metabolism, such as red cell protoporphyria. Measurable increases in the EP level take 2 to 3 weeks of sustained iron deficiency, and a return to normal with iron therapy takes even longer. Therefore, an EP level is a more stable measurement of chronic iron deficiency, in contrast to the serum iron level or transferrin saturation, which can be depressed within hours during a transient infection. In addition, EP levels indicate an imbalance between the iron supply and the need of the individual red cell. This means that, in addition to the iron-deficient erythropoiesis of true iron deficiency and inflammation, a high EP level can indicate a relative iron deficiency, in which plasma iron supply would be adequate for normal needs but insufficient to meet the increased requirement of a proliferative marrow.

Chapter

3 Differential Diagnosis of Anemia

The cause of an anemia may be apparent from the history and physical examination. For example, when a patient is actively bleeding, a fall in the hemoglobin level is predictable, whereas bleeding over a longer period can result in an iron deficiency anemia. The presence of symptoms and signs of inflammation or a chronic systemic illness can be enough to explain a moderately severe hypoproliferative anemia. Splenomegaly may suggest the possibility of a hemolytic anemia; drug exposure can cause an aplastic anemia. However, the mere coexistence of anemia with another illness does not reliably pinpoint the cause. Accurate determination of the cause of an anemia requires a full laboratory evaluation with studies of blood and marrow morphology, red blood cell production, iron supply, and, in some cases, red blood cell destruction.

Several algorithms have been proposed to organize the approach to anemia diagnosis. Changes in red blood cell morphology can be used to classify anemias according to whether they are normocytic, microcytic, or macrocytic. This approach is made somewhat more sensitive by using the RDW to make the distinction between microcytosis secondary to iron deficiency and that caused by thalassemia (Fig. 37). The morphologic approach to anemia diagnosis is not reliable, however, if the anemia is relatively mild or of recent onset, since the stimulus required to produce abnormal cells is not present.

Another approach to the classification of an anemia uses the reticulocyte production index and the CBC to classify the anemia as resulting from one of three functional defects: a failure in red cell production (a hypoproliferative anemia); a disturbance in cell de-

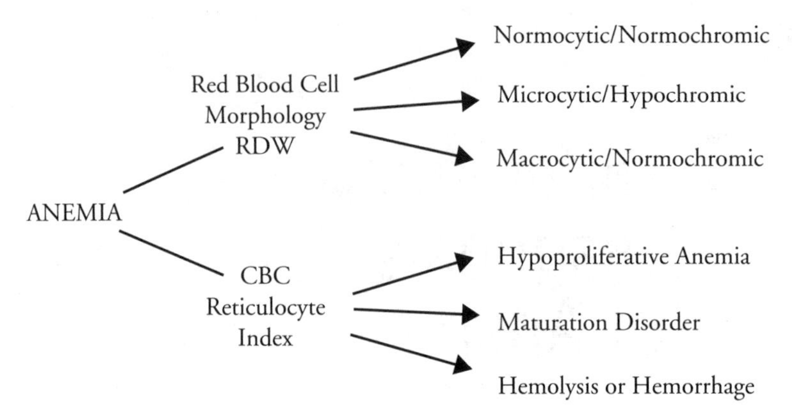

Figure 37. Approach to the diagnosis of an anemia. As a first step, the complete blood count (CBC), reticulocyte index, RDW, and red blood cell morphology can be used to broadly categorize the type of anemia. The CBC and reticulocyte index are emphasized in the functional classification approach.

velopment (a maturation disorder); or an increase in destruction (a hemolytic or hemorrhagic anemia). Hypoproliferative anemias demonstrate a low reticulocyte index, together with little or no change in red cell morphology. This can result from a failure in any one of the major components involved in red cell production, including the response of erythropoietin synthesis to the anemia, the proliferation of marrow erythroid precursors, or the supply of iron (Fig. 38). In contrast, maturation disorders demonstrate a low reticulocyte production index, reflecting an ineffective marrow response, and a highly abnormal red cell morphology (macrocytosis or microcytosis). Patients with a hemolytic or hemorrhagic anemia may be distinguished on the basis of their increased reticulocyte production index (greater than three times basal) and, in some cases, changes in red cell morphology.

The functional classification of anemia also provides a useful starting point in organizing other laboratory studies needed to complete the differential diagnosis (Fig. 39). For example, the diagnosis of a hypoproliferative anemia requires a bone marrow aspirate and biopsy, iron supply studies, and measurements of both renal and endocrine function to differentiate between a failed erythropoietin response, marrow damage, and iron deficiency. Important studies in the differential diagnosis of a maturation disorder include measurements of vitamin levels and the structural characteristics of hemoglobin. The differential diagnosis of a he-

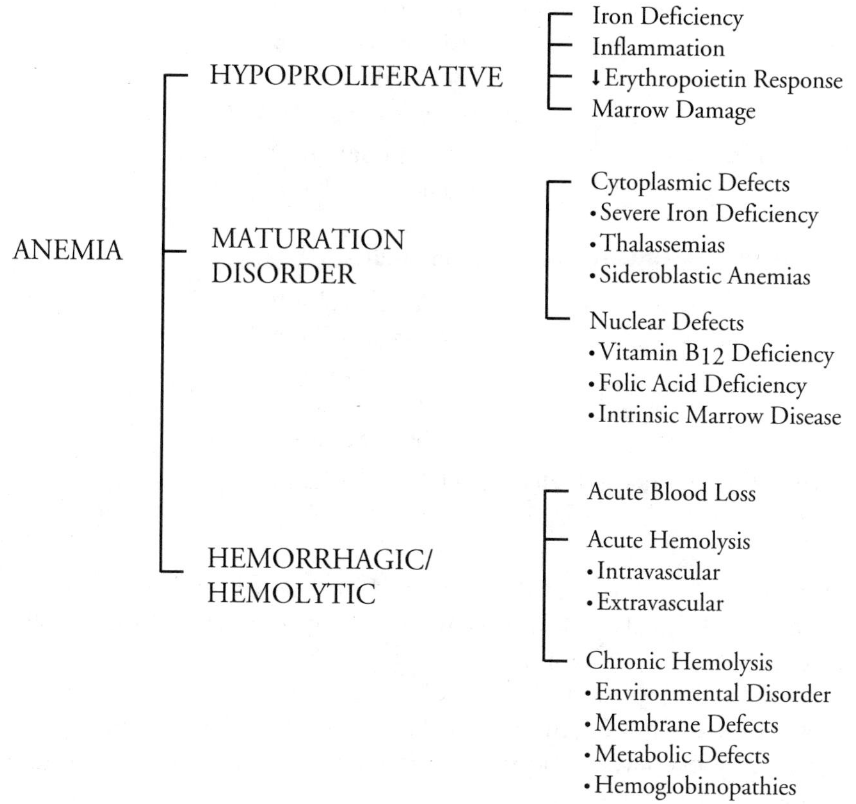

Figure 38. The full functional classification of anemia. The classification and diagnosis of an anemia is based on clinically recognizable abnormalities in erythron function. Initially, anemias can be subdivided as hypoproliferative, a maturation disorder, and hemorrhagic or hemolytic. Each of these categories can then be further analyzed to reach a more specific diagnosis.

molytic anemia requires additional laboratory studies to look for membrane and metabolic abnormalities.

The Hypoproliferative Anemias

The typical erythropoietic profile of a hypoproliferative anemia is summarized in Table 7. The combination of normocytic, normochromic morphology and a reticulocyte index that is inappropriately low for the severity of the anemia (generally less than 2) are the key findings that suggest this group of disorders. Additional evidence of the hypoproliferative nature of the anemia includes a

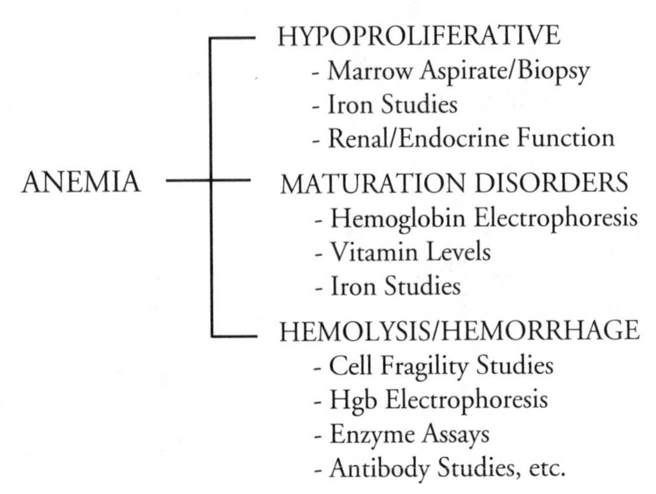

Figure 39. The functional classification of anemia. This approach is helpful for organizing the laboratory studies needed in the diagnosis of the individual anemias.

marrow E:G ratio of less than 1:2 and normal or decreased bilirubin and LDH levels. Laboratory measurements, such as iron supply studies, serum erythropoietin levels, and chromosome analysis, vary according to the individual disease state. Therefore, although these tests are not useful in classifying an anemia as hypoproliferative, they do play a role in the differential diagnosis of the individual disease states.

Table 7 • **Hypoproliferative Anemia: Erythropoietic Profile**

Smear/Red Cell Indices
Normocytic, normochromic with minimal anisocytosis or poikilocytosis; MCV = 85–95 fL

Reticulocyte Index
Less than 2.0
Polychromasia may or may not be present depending on the etiology. (It is present with marrow damage and iron deficiency and absent with renal failure and inflammation.)

Marrow E:G Ratio
Less than 1:2
Morphology may be diagnostic for marrow damage (fibrosis, tumor invasion, leukemia).

Indirect Bilirubin Levels
Decreased to normal

Most anemias encountered in clinical practice are of the hypoproliferative type (Fig. 40). They are most often associated with an acute or chronic inflammatory disease state. Suppression of early stem-cell proliferation (marrow damage) is seen much less frequently. The diagnosis of iron deficiency and inflammatory anemias should be relatively straightforward, requiring little more than measurements of iron supply. Therefore, these conditions will be discussed first as a prototype.

Iron Deficiency Anemia

Iron deficiency is defined as a state of negative iron balance in which iron supply is inadequate to meet the iron requirements of the erythron and other body tissues.[51] The risk of developing iron deficiency is influenced by age, sex, and medical status. Body iron content at birth is about 300 mg and increases thereafter to meet the requirements of growth, eventually stabilizing at 2 to 4 g in adulthood (Table 8). This complement of iron is maintained in men by the absorption of about 1 mg per day, which represents approximately 7% of available dietary iron (Fig. 41). The premenopausal woman has a greater requirement because of an average menstrual loss of 0.5 mg per day (range is 0.2 to 2.0 mg per day) in addition to the other physiologic losses. This means that a woman's iron absorption under basal conditions must approach

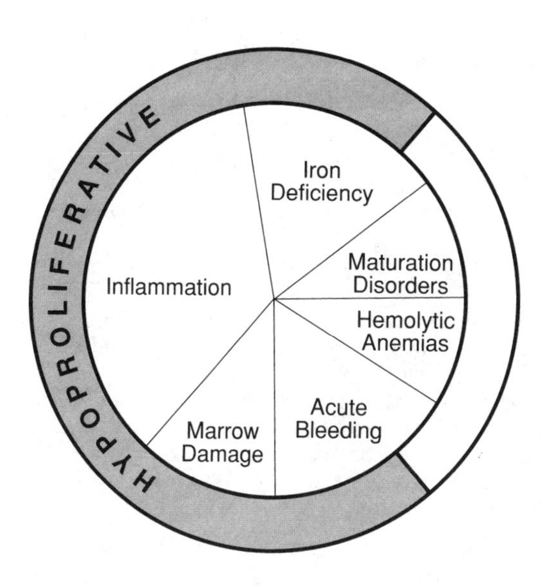

Figure 40. Relative frequency of types of anemia in practice. The majority of anemias in practice are hypoproliferative; inflammation, iron deficiency and acute bleeding are the most common disorders. Disorders of marrow stroma and/or stem cells are less common, as are the maturation disorders and hemolytic anemias.

Table 8 • **Body Iron Content (mg)**

	Body Weight	
	80 kg ♂	65 kg ♀
Erythron iron	2400	1700
Iron stores	1000 ± 200	300 ± 300
Myoglobin/tissue iron	400	300
Transferrin	6	4
Total body iron (g)	4	2.5

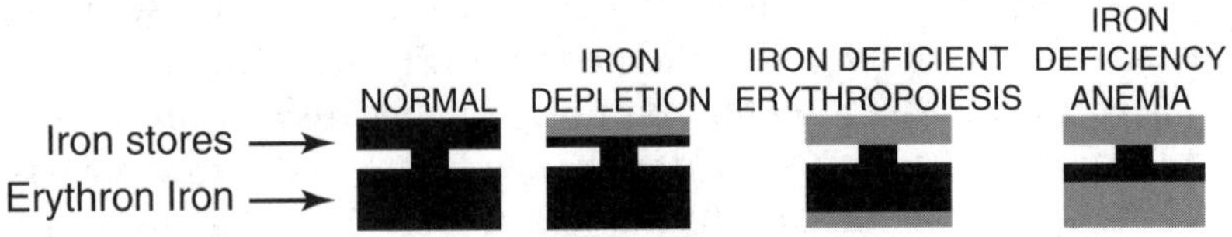

	NORMAL	IRON DEPLETION	IRON DEFICIENT ERYTHROPOIESIS	IRON DEFICIENCY ANEMIA
RE Marrow Fe	2 - 3 +	0 - 1 +	0	0
Transferrin IBC (μg/dL)	330 ±30	360	390	410
Serum Ferritin (μg/mL)	100 ±60	20	10	<10
Iron Absorption (%)	5 - 10	10 - 15	10 - 20	10 - 20
Serum Iron (μg/dL)	115 ±50	115	<60	<40
Transferrin Saturation (%)	35 ±15	30	<15	<10
Sideroblasts (%)	40 - 60	40 - 60	<10	<10
RBC Protoporphyrin (μg/dL RBC)	30	30	100	200
Erythrocytes	Normal	Normal	Normal	Micro/Hypo

Figure 41. Sequential changes in the development of iron deficiency. Indicators of iron store depletion include the visual inspection of marrow reticuloendothelial iron stores, the total iron-binding capacity (TIBC), the serum ferritin level, and the percent of iron absorbed from an oral iron test dose. With the onset of iron deficient erythropoiesis, the serum iron, percent saturation of transferrin, percent sideroblasts observed on the marrow iron stain, and erythrocyte protoporphyrin become abnormal. At this point, anemia appears. When anemia has been present for some time, red cells become microcytic and hypochromic.

13% of the iron in her diet, nearly twice that of a man. Based on these differences in dietary iron content, iron absorption, and iron losses, the amount of iron stores in men is usually between 600 and 1200 mg as compared to 100 to 400 mg in women.

There are certain critical periods in iron balance.[51] After the initial 4 to 6 months of life, during which time the infant relies on iron stores accumulated during gestation, iron balance becomes precarious because of a limited iron intake at a time when growth requirements are great. Without iron supplements, iron deficiency in infants is common. In developing countries, 20% to 40% of infants may be iron deficient; in the United States, the prevalence is much less mostly because of the use of iron-fortified formulas. Later in childhood, the increasing caloric intake, carrying with it proportionately more iron, usually provides an adequate iron supply. However, continued rapid growth through adolescence prevents any appreciable accumulation of iron stores.

The nature of the individual's diet also influences iron balance. In the Western diet, there are about 6 mg of iron per 1000 calories of food, so that caloric intake is relevant. Even more important is the bioavailability of the dietary iron. The most available form is meat-derived heme iron, the absorption of which is unaffected by the general composition of the diet. Non-heme iron is less available, and its absorption is dependent on facilitators such as ascorbic acid and meat protein. A well-balanced Western diet contains 15 to 20 mg of iron, so that the iron-depleted individual can absorb as much as 3 to 4 mg per day or 20% to 25% of the available iron. People of low economic status who cannot afford meat and fruit may absorb no more than 1 to 2 mg of the iron from their diet. The phytates contained in whole grains and vegetables also inhibit iron absorption.

An imbalance between dietary iron intake and physiologic iron losses is the most common cause of iron deficiency in children and young women. This is not true for men or postmenopausal women, in whom dietary iron balance is favorable. The appearance of iron deficiency in these adults almost invariably means pathologic blood loss (Table 9). Acute and chronic blood loss secondary to gastrointestinal disease, repeated blood donation, surgical blood losses, intravascular hemolysis, or abnormal gynecologic bleeding can rapidly deplete iron stores and lead to iron-deficient erythropoiesis and anemia. When there is no evident blood loss, iron malabsorption secondary to a gastrointestinal disease process may be the cause.

Table 9 • **Causes of Iron Deficiency**

Nutritional	**Malabsorption**
Infant	Postgastrectomy
Menstruating or pregnant woman	Sprue
Bleeding	**Hemosiderin Loss**
Uterine	Pulmonary siderosis
Gastrointestinal	Hemosiderinuria
Traumatic	

As a part of the differential diagnosis of iron-deficient erythropoiesis, the impact of acute and chronic inflammation on iron supply must also be considered. Inflammatory cytokines not only suppress erythropoietin and stem-cell proliferation but also interfere with the delivery of iron from reticuloendothelial stores to transferrin for transport to the erythroid marrow. This reduces the plasma iron and can, over time, result in a microcytic, hypochromic anemia.

The diagnosis of an iron-deficiency state depends on a number of laboratory measurements (see Fig. 41). Before the development of an anemia, *negative iron balance* can be detected from the depletion of reticuloendothelial stores, as seen on the Prussian blue stain of a marrow aspirate.[50] Negative iron balance can also be identified from the serum ferritin level, which falls as stores are depleted.[52] Normal values used in determining iron depletion are different for men and women. The amount of iron stores in a normal man is usually between 600 and 1200 mg (2 to 3+ visible marrow iron stores), as compared to 100 to 400 mg (0 to 1+ visible marrow iron stores) in the normal woman. An accurate diagnosis of negative iron balance must recognize this difference.

Studies of iron absorption and plasma transferrin concentration, or TIBC, can also suggest negative iron balance. However, iron absorption is difficult to quantitate and highly variable, whereas the expected increase in transferrin (TIBC) may be prevented by any factor that depresses protein production. Other measurements of iron supply (see Fig. 41) are normal. This includes the plasma iron level, the percent saturation of transferrin, the percentage of sideroblasts in the marrow, and the red cell protoporphyrin level. Moreover, the hemoglobin level is within normal limits and there is no change in red cell morphology.

Once iron loss is sufficient to exhaust reticuloendothelial stores, iron supply to the erythron decreases. At this point, the patient has *iron-deficient erythropoiesis,* even though a significant anemia or change in red cell morphology is not yet apparent. As shown in Figure 41, all tests of iron supply now demonstrate a defect in iron delivery. As iron stores are exhausted, the serum ferritin level falls below 12 μg/L and the plasma iron falls below 50 μg/dL. More important, the percent saturation falls below 16%. This reduces the supply of iron to erythroid precursors, as revealed by a decrease in the number of sideroblasts on the marrow iron stain and an increase in the red cell protoporphyrin to levels greater than 70 μg/dL of red blood cells.

Finally, as iron loss continues, the complete picture of iron deficiency anemia appears. In addition to the abnormal measurements of iron supply, the hemoglobin concentration falls, and the circulating red cells become microcytic and hypochromic. The degrees of microcytosis and hypochromia correlate with the severity of the anemia and its duration. Marked reductions in the MCV to levels below 70 fL appear only when the hemoglobin falls to levels below 10 g/dL for weeks or months. Once microcytosis and hypochromia are present, there should be little difficulty in diagnosing iron deficiency as the cause of the anemia. All of the measurements of iron supply will be highly abnormal.

It is important clinically to understand the stages of iron deficiency and how they affect the type of anemia observed in any patient. The sequence of iron depletion and anemia in patients with acute blood loss is a good example. Sudden hemorrhage in an otherwise normal individual acutely depresses the hemoglobin level. Because iron stores are initially sufficient to support at least a modest increase in the production of new red blood cells, the first response is an increase in the reticulocyte production index, together with a normal or even slightly elevated MCV caused by the release of shift cells (marrow reticulocytes). The magnitude of the reticulocyte response reflects the severity of the anemia and the size and availability of the patient's reticuloendothelial iron stores. In general, men do better than women. Men have more iron stores to draw on and are more responsive to a moderately severe anemia because of their higher basal hematocrit level.

With continued blood loss, iron stores are exhausted and iron supply to the marrow falls below the level necessary to support even basal levels of red cell production. The reticulocyte production index falls, resulting in the hypoproliferative anemia picture of iron-deficient erythropoiesis. At this point, red cell morphology is still normal. Microcytosis and hypochromia appear only after further iron loss and passage of some weeks or months. Other features of severe iron deficiency, including soreness of the mouth, difficulty swallowing, abnormal cravings for certain foods or ice (pica), and spooning of the nails appear only when tissue iron deficiency is extreme. Therefore, the findings that support the diagnosis of iron deficiency vary according to the clinical picture, and the clinician needs to be skilled in the use of the laboratory measurements of iron supply so as not to be fooled.

Iron deficiency may also be anticipated because of the nature of the patient's disease process or other screening tests (see Table 9). External bleeding (recurrent epistaxis, bleeding hemorrhoids, prolonged menses, frequent blood donations, and so on) is usually obvious from the history. Occult blood loss from the gastrointestinal tract may be detected by guaiac-testing one or more stool samples. Even repeated negative guaiac testing is not foolproof, however, and a study of the gastrointestinal tract by x-ray or endoscopy in symptomatic patients may reveal a potential site of acute or chronic blood loss. Other situations in which iron deficiency is likely include malabsorption (sprue, Crohn's disease, gastrectomy, or small-bowel surgery), intravascular hemolysis with hemosiderinuria (paroxysmal nocturnal hemoglobinuria), and pulmonary hemosiderosis. The diagnosis of a malabsorption defect involves studies of both the anatomy and function of the small bowel.

Iron absorption can be clinically evaluated using an oral iron tolerance test. After an oral dose of 50 mg of iron as ferrous sulfate, plasma iron levels are determined every 30 minutes over the next several hours. A rapid rise in the plasma iron level of over 50 to 100 μg/dL indicates normal absorption. The patient who has undergone gastric surgery represents a special situation in which absorption of iron salts may be intact but food iron absorption is impaired by a lack of gastric acidification, a rapid transit time, or both. Such patients can become severely iron deficient unless they supplement their diets with medicinal iron.

Patients who have repeated episodes of intravascular hemolysis can become iron deficient because of losses of iron into the urine. Hemosiderinuria may be detected by examination of urinary sediment stained with Prussian blue. The presence of iron-staining particles within tubular cells is diagnostic. One cause of this type of iron loss is paroxysmal nocturnal hemoglobinuria, an acquired defect in the red cell membrane that can be detected using a sugar water screening test or Ham's acid hemolysis test. Pulmonary hemosiderosis may be detected by the demonstration of iron-staining particles in sputum macrophages. Severe pulmonary hemosiderosis is associated with the appearance of visible iron deposits on the chest x-ray.

Unusual forms of tissue iron deficiency are seen in patients with copper deficiency and certain types of metal toxicity, those on renal dialysis, and those with certain industrial exposures. Microcytic anemias are associated with cadmium, lead, and aluminum toxicity. This may be related to the iron status of the affected subject because the absorption of these metals is increased in the iron-depleted individual. These metals also directly affect hemoglobin synthesis.

Only a few diseases need to be considered in the differential diagnosis of iron deficiency. Chronic inflammatory diseases typically exhibit a hypoproliferative anemia that may or may not be mildly microcytic and hypochromic. However, the pattern of iron supply measurements should distinguish the anemia of inflammation (anemia of chronic disease) from true iron deficiency (Table 10). Although the plasma iron falls to low levels with inflammation, the TIBC level (less than 300 μg/dL) and serum ferritin level (greater than 20 μg/dL) differ greatly from that seen with true iron deficiency.

When patients present with a moderate to severe microcytic, hypochromic anemia, they need to be distinguished from patients with thalassemia (a globin-production defect) or a sideroblastic

Table 10 • **Differential Diagnosis of Iron Deficiency**

	Plasma Iron/TIBC	Saturation	Ferritin	Iron Stores
	(μg/dL)	(%)	(μg/L)	
Iron deficiency	<30/>350	5–12	<12	Absent
Inflammation	<30/<300	10–20	20–200	Present
Thalassemia	>50/Normal	30–90	20–1000	Present
Ring sideroblastic anemias	>50/Normal	30–90	20–1000	Present

anemia (an inherited or acquired defect in mitochondrial function). Here again, the iron supply studies usually provide a differential diagnosis. The thalassemias and sideroblastic anemias show normal to increased plasma iron, normal TIBC, and increased serum ferritin levels (see Table 10). Iron deficiency and thalassemia can occur together. This is relatively common in children with β thalassemia minor and nutritional iron deficiency. Iron studies will confirm the iron-deficiency state, and studies of hemoglobin structure will be needed to confirm the globin production defect.

Acute and Chronic Inflammation

Anemia is a common component of inflammatory states. The presence of an inflammatory anemia can even help in the diagnosis of the patient's primary disease state. For example, the appearance of a hypoproliferative anemia can be an early clue to an underlying collagen vascular disease, such as polymyalgia rheumatica or temporal arthritis, even before other symptoms and signs are well developed. In patients with rheumatoid arthritis, the severity of the hypoproliferative anemia can serve as an indicator of disease activity. Flare-ups of the patient's arthritis are accompanied by worsening anemia.

Acute infection or inflammation produces a hypoproliferative anemia within a matter of days. This reflects the systemic effects of infection and the impact of inflammatory cytokines on both red cell survival and the new red cell production. The initial fall in the patient's hematocrit during the first few days of an infection is caused by a sudden loss from circulation of red blood cells that are near the end of their natural life span. Older red blood cells cannot maintain their normal pliability and structure in the face of a systemic inflammatory reaction and are rapidly removed by the reticuloendothelial system.

The principal mechanism responsible for the anemia of acute or chronic inflammation is, however, not the reduction in red cell life span.[53] Rather, it is the impact of inflammatory cytokines on the erythropoietin response (resulting in a decrease in marrow stimulation), the level of iron supply, and the proliferative response of erythroid marrow precursors (Fig. 42). Several cytokines have been implicated, including tumor necrosis factor (TNF), interleukin-1 (IL-1), interferon-γ(IFN-γ) and interferon-β(IFN-β), and neoptrin. The increase in TNF levels in patients with neoplasms and bacterial infections directly suppresses the erythropoietin response and re-

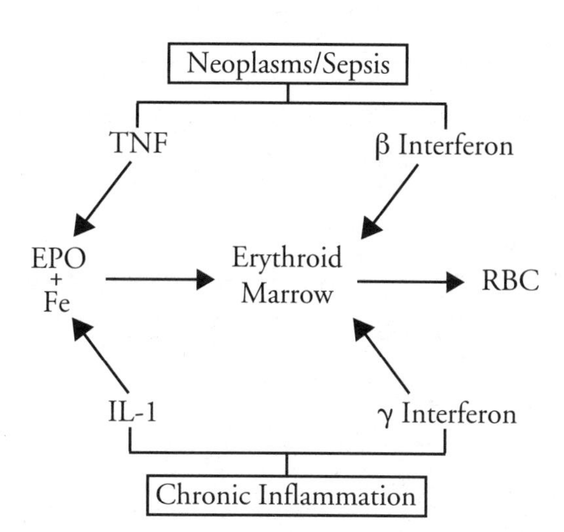

Figure 42. Inflammatory cytokine suppression of erythropoietin (EPO), iron supply, and erythroid marrow proliferation. Cytokines implicated in the inflammatory anemias include tumor necrosis factor (TNF), γ and β interferons (IFN-γ and IFN-β), interleukin-1 (IL-1), and neoptrin.

duces the plasma iron by blocking the release of iron from reticuloendothelial stores.[54] TNF also induces IFN-β to suppress the growth of erythroid marrow precursors. In patients with chronic inflammatory states, such as rheumtoid arthritis, IL-1 suppresses erythropoietin, iron supply, and (via IFN-γ and/or neoptrin) marrow erythroid precursor growth.

The typical erythropoietic profile of the anemia of acute and chronic inflammation is summarized in Table 11. For the patient with an acute infection, the anemia is generally mild, with a hematocrit greater than 30%, and the red cell morphology is normocytic and normochromic. The reticulocyte production index is typically less than 2 and, because erythropoietin is suppressed, polychromasia is absent. With chronic inflammatory states, the anemia takes on more of the manifestations of iron deficiency. Hematocrit levels can fall below 30% and the smear may be mildly microcytic (MCV of 75 to 85 fL) and slightly hypochromic. Marrow production indices are still hypoproliferative, however, The reticulocyte index is less than 2, and polychromasia is not observed. If a marrow aspirate is performed, the marrow E:G ratio is less than 1:2.

The clinical feature that distinguishes the anemia of acute and chronic inflammation from the other hypoproliferative anemias is the pattern of the iron supply measurements. Cytokine interference with the release of iron from marrow reticuloendothelial stores to transferrin provides a unique diagnostic marker. As summarized in

Table 11 • **Inflammatory Anemia: Erythropoietic Profile**

Smear/Red Cell Indices
Normocytic/normochromic to mildly microcytic/hypochromic (chronic inflammation) with little or no anisocytosis
Reticulocyte Index
Less than 2.0
Polychromasia absent
Marrow E:G Ratio
Less than 1:2, especially with acute bacterial infections with granulocytosis
Indirect Bilirubin/LDH levels
Normal

Table 10, the plasma iron falls to levels less than 30 μg/dL and the TIBC decreases, which produces a percent saturation of transferrin of 10% to 20%. This fall in iron supply occurs despite the presence of reticuloendothelial-cell iron stores, as measured by either the Prussian blue stain of a marrow aspirate or the serum ferritin level. In prolonged chronic inflammatory disease states, marrow iron stores even appear to increase, with reticuloendothelial cells showing prominent deposits of large iron granules. The serum ferritin also increases, reaching levels in excess of 100 to 200 μg/L.

The inflammatory pattern of iron supply measurements is distinctly different from that seen with the iron deficiency of iron loss (see Table 10). Therefore, the differential diagnosis in patients with an uncomplicated anemia of either type should not be difficult. An anemia caused by multiple factors can be a diagnostic challenge, however. For example, patients with alcoholic liver disease often present with combined blood loss, folic acid deficiency, and inflammatory liver disease. In this situation, the inflammatory pattern of iron supply and hypoproliferative red cell production may not be apparent until the folic acid and iron deficiency are treated. Patients with rheumatoid arthritis often demonstrate a combination of true iron deficiency caused by bleeding secondary to aspirin ingestion and inflammation from their primary disease. In this setting, iron stores may be absent on the marrow aspirate, yet the serum ferritin level, instead of being less than 12 μg/L as with iron deficiency or greater than 100 μg/L as seen with inflammation, falls somewhere in between.

Decreased Erythropoietin Response

Hypoproliferative anemias are common in patients with severe renal disease, protein deprivation, and endocrine deficiency states. Loss of renal tissue and renal function are almost always accompanied by a partial failure of erythropoietin regulation.[55] The defect is not a complete loss of all erythropoietin production, but rather an inability to increase the production of erythropoietin by an amount appropriate to the level of anemia (Fig. 43). The severity of the anemia usually mirrors the severity of the renal failure. With increasing uremia, toxic products in circulation shorten the life span of red blood cells to produce a worsening anemia. However, it is the underlying failure of the erythropoietin response that determines the hyproproliferative nature of the anemia.

Typical of a hypoproliferative state, renal anemia is characterized by a normal MCV and normocytic, normochromic morphology on the blood smear. Sometimes individual cell abnormalities, including targeting or the appearance of burr or helmet-shaped cells, are seen, especially in patients with severe uremia. Even with severe renal failure, the hemoglobin level usually does not drop below 7 g/dL, and the anemia is fairly well tolerated. The reticulocyte index is typically less than 2, and the bilirubin level is less than 0.4 mg/dL. Examination of the marrow reveals a degree of proliferation far lower than expected for the degree of anemia. As long as there has not been blood loss or iron overload from repeated transfusion, iron status is normal. This includes a normal plasma iron count, TIBC, and serum ferritin level.

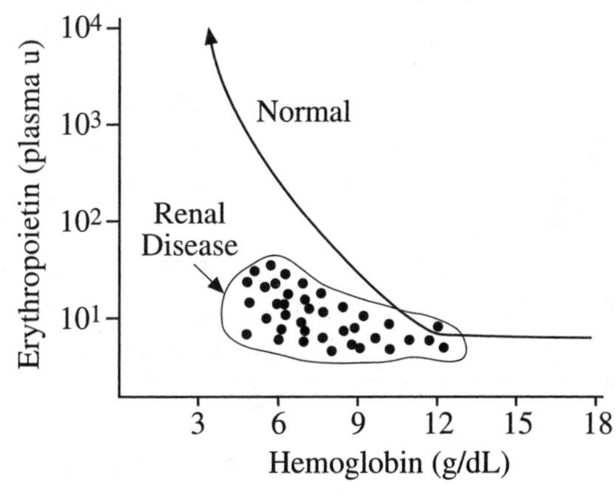

Figure 43. Plasma erythropoietin levels in patients with renal disease. The failure of the erythropoietin response to anemia is the principal cause of the hypoproliferative anemia associated with loss of renal function.

For predialysis patients, the severity of the anemia roughly correlates with the rise in blood urea nitrogen (BUN) and/or creatinine levels. In some clinical situations, however, the correlation between excretory and endocrine function of the kidney is lost. For example, arteriolar lesions of the hemolytic-uremic syndrome do not interfere with the erythropoietin response. These patients show severe excretory malfunction, but they are able to respond with an appropriate increase in red cell production. At the opposite extreme, patients with diabetic nephropathy can demonstrate a degree of anemia greater than that expected for the level of nitrogen retention. This would suggest that the diabetic renal lesion has an early effect on the tubular interstitial cells responsible for erythropoietin secretion.

A decrease in erythropoietin stimulation is also seen with endocrine deficiencies or altered hormonal states.[56] With a single hormone deficiency (testosterone or thyroid hormone), the hemoglobin falls by 1 to 4 g/dL. Castration or estrogen administration to a man reduces the hemoglobin level by 1 or 2 g/dL. Pituitary deficiency, with its effect on multiple target organs, can depress the hemoglobin level by as much as 3 to 6 g/dL. In all cases, the anemia is typically hypoproliferative, with normal red blood cell indices and normocytic, normochromic morphology. With adrenal or pituitary insufficiency, the white blood cell count may be moderately decreased to 3000 to 5000/μL and may fall further with inflammation instead of increasing. There may also be a modest eosinophilia.

The mechanisms whereby various hormones affect blood production are not well defined. The anemia associated with thyroid hormone deficiency can be largely attributed to the decrease in metabolic rate. On the other hand, androgenic hormones have a stimulating effect on erythropoietin production and erythropoietin action at the stem-cell level. A moderate decrease in hemoglobin level may be seen in patients with protein deprivation.[57] Like the endocrine effect, this is attributed to a decrease in erythropoietin drive based on a decrease in metabolic rate. This type of anemia is usually very mild, a 1 to 3 g/dL deficit in hemoglobin with a somewhat greater reduction of red cell mass caused by an associated decrease in total blood volume. Red blood cell indices are normal, and the blood smear is normocytic and normochromic.

When a patient with an endocrine abnormality or protein deprivation demonstrates a more severe anemia, a hemoglobin less than 10 g/dL, it is important to look for other causes of anemia, particularly iron- and vitamin-deficiency states. Patients with severe hypothyroidism for a prolonged period can develop nutritional deficiencies, including iron and vitamin malabsorption. Protein deprivation is often accompanied by a poor intake of essential vitamins and iron or a marked defect in small-bowel absorption. With refeeding, the patient with severe marasmus (protein-calorie malnutrition) shows a temporary drop in hemoglobin caused by a rapid expansion of plasma volume. This may also be seen with hormonal repletion in patients with Addison's disease. Generally, the response of the hemoglobin level is slow; full correction of the anemia takes many weeks or even months.

Rarely, an inherited hemoglobinopathy or metabolic deficiency of the red blood cell results in a modest anemia because of a decreased affinity of hemoglobin for oxygen.[58] With decreased oxygen affinity, a normal amount of oxygen is delivered to tissues despite the reduced amount of hemoglobin. This phenomenon may be detected by measuring the red cell P_{50}. The most common example of this situation is pyruvate kinase deficiency, in which the associated increase in 2, 3-DPG increases oxygen release. This makes it possible for the patient to reach a compensated level of oxygen delivery at hemoglobin levels of 9 to 11 g/dL.

Hyperthyroid patients can demonstrate a mild hypoproliferative anemia. The explanation for this is not clear. Very high levels of parathyroid hormone may directly inhibit stem-cell proliferation or, because of the development of renal calcification, interfere with erythropoietin secretion.

The anemia observed during the second and third trimesters of pregnancy most often reflects the physiologic changes in the patient's blood volume. A modest reduction in the hemoglobin to levels of 10 to 11 g/dL is expected. At this level, most women show an actual increase in the red cell mass but a greater increase in their plasma volume to produce the modest reduction in hemoglobin level. However, iron and folic acid deficiency are extremely common in pregnancy because of the increased requirements of the fetus. Both iron and vitamin supplements should be given throughout pregnancy to prevent a deficiency state. If the hemoglobin level falls below 10 g/dL during pregnancy, the patient should be carefully evaluated for iron or folate deficiency, an inflammatory condition, and/or renal disease.

Marrow Damage

Significant damage to the marrow structure or the erythroid precursor pool results in a hypoproliferative anemia. Most often, all marrow cell elements are involved, so that patients present with varying combinations of anemia, leukopenia, and thrombocytopenia. Rarely, only the erythron-committed stem cells are involved, resulting in pure red-cell aplasia.[59] Mild marrow-damage anemias are typically seen with solid tumor infiltrations of the marrow and with drug toxicities. Patients with acute leukemia or severe aplastic anemia present with more severe anemia, a marked thrombocytopenia, and an absence of normal granulocytes in circulation, with or without the appearance of large numbers of malignant cells.

The cause of a marrow-damage anemia may be obvious from the clinical history. Patients undergoing multidrug cancer chemotherapy predictably develop an anemia or pancytopenia. The severity of the anemia and pancytopenia can usually be anticipated on the basis of the type and number of chemotherapy drugs used. The cause of the anemia is also apparent in patients who present with a primary hematopoietic malignancy. Infiltration of the marrow by a leukemic cell line will effectively halt normal red cell production. The presence of excessive blasts in circulation or in marrow makes the diagnosis. Marrow-damage anemia associated with the more gradual onset of a myeloproliferative disorder can be a greater diagnostic challenge. Marrow morphology, including a marrow biopsy to evaluate overall cellularity, is important to identify the cause of the anemia. This is also true for patients who present with a severe aplastic anemia or a disorder of marrow architecture (stromal disease).

The erythropoietic profile of a moderately severe marrow-damage anemia is summarized in Table 12. Red blood cell morphology is generally normocytic and normochromic, and the reticulocyte production index is below that expected for the severity of the anemia, which is clearly hypoproliferative. Examination of the peripheral blood smear may reveal a number of clues to the nature of the disorder, including the appearance of large numbers of abnormal white blood cell precursors (blasts), nucleated red blood cells, and poikilocytic red blood cells, especially teardrop cells. Similarly, the marrow E:G ratio and morphology frequently provide diagnostic information.

The pattern of iron supply studies reflects the severity of the

Table 12 • **Marrow-Damage Anemia: Erythropoietic Profile**

Smear/Red Cell Indices
- Normocytic, normochromic. Abnormal cell forms (blasts, nucleated red blood cells, teardrops) may be present.
- MCV = 85–95 fL

Reticulocyte Index
- Less than 2.0
- Polychromasia usually present, unless aplasia is severe

Marrow E:G Ratio
- Less than 1:3
- Morphology often diagnostic (e.g., hypocellular, tumor infiltration, leukemia, fibrosis)

Iron Supply Studies
- Serum iron normal to increased
- TIBC normal
- Ferritin normal to increased

marrow damage. The serum iron, TIBC, and serum ferritin levels are all normal in patients with mild to moderate marrow damage. Severe damage or aplasia is associated with a rise in the serum iron to full saturation of the iron-binding capacity. In addition, the serum ferritin level rises, reflecting an accumulation of iron in the reticuloendothelial cells. This iron supply pattern reflects the loss of the normal pathway of iron flow to marrow red blood cell precursors.

The combination of normocytic, normochromic red blood cell morphology, abnormal bone marrow morphology, and normal or elevated serum iron and ferritin levels readily distinguishes a marrow-damage anemia from hypoproliferative anemias secondary to iron deficiency, an inflammatory illness, or a failure of erythropoietin response. The differential diagnosis of the cause of the marrow damage is a greater challenge. If there is a clear association with an inciting agent such as chemotherapy, the diagnosis is easy, and the extensive evaluation is unnecessary. When this is not the case, a number of causes of stem-cell or structural damage to the marrow must be considered (Table 13).

In addition to the known effects of radiation and chemotherapy drugs on marrow stem cells, a number of other drugs have been

Table 13 • **Causes of Marrow Damage**

Causes of Marrow Damage
Stem-Cell Damage
Chemotherapy or radiotherapy
Drugs—e.g., antibiotics, antidepressants
Chemical agents—e.g., solvents, heavy metals
Bacterial or viral infections
Stem-cell malignancies—e.g., leukemia
Aplastic anemia
Structural Damage
Radiation
Metastatic malignancies
Myelofibrosis
Granulomatous diseases—e.g., tuberculosis
Gaucher's disease
Autoimmune or Unknown
Infections—e.g., hepatitis
Rheumatologic disease—e.g., SLE
Aplastic anemia
Pure red blood cell aplasia
Graft-versus-host disease
Congenital anemias

implicated in the development of a severe (at times irreversible) stem-cell damage. Drugs commonly associated with severe aplasia include several antibiotics, the tricyclic antidepressants, the anti-inflammatory drug phenylbutazone, the thiouracil-derivative drugs used in treating thyroid disease, and some diuretics.[60] However, it is good clinical practice to consider any medication as suspect and, if possible, discontinue its use. If this results in an improvement in the patient's anemia or pancytopenia, it provides at least circumstantial evidence of drug effect.

Occupational exposure to chemicals, especially those that contain benzene or benzene derivatives, can cause aplastic anemia, acute leukemia, or both. As with drugs, any exposure to solvents, insecticides, or heavy metals should be considered as a possible cause of marrow stem-cell damage.

High-energy radiation is also recognized for its predictable marrow toxicity. When patients receive whole-body radiation, the toxic effect on the marrow will be dose-dependent. Transient falls in the blood counts are seen at total body doses of less than 250 rads, and irreversible loss of stem cells occurs at higher doses.

Marrow-damage anemias are also seen with bacterial and viral infections.[59] Patients with miliary tuberculosis can develop extensive granuloma formation and marrow fibrosis, resulting in a worsening anemia and even pancytopenia. When granulomas are present in a marrow biopsy or aspirate specimen, mycobacteria may be identified using an acid-fast stain. Viral illnesses, including viral hepatitis, human immunodeficiency virus (HIV), Epstein-Barr virus (EBV), and parvovirus, can result in a severe aplastic anemia or, in the case of parvovirus, pure red cell aplasia. The mechanism involved is still unclear but appears to involve an autoimmune process. Parvovirus infection in patients with congenital hemolytic anemias is now recognized as a cause of transient aplastic crises.

Malignancies (both primary hematopoietic malignancies and metastatic tumors) are another common cause of moderate to severe marrow-damage anemia. In the case of metastatic solid tumors, the focal growth of tumor cells in the marrow sufficiently disrupts the marrow structure to compromise red cell precursor growth. In addition, tumors can secrete cytokines that suppress stem-cell proliferation and stimulate fibrosis. Hematopoietic malignancies invariably produce a severe marrow-damage anemia. In this situation, the growth advantage of the hematopoietic tumor cell displaces normal red cell production.

Patients can also develop a marrow-damage anemia or aplastic anemia secondary to an autoimmune disease process. The most dramatic examples of this are the young individuals who suddenly develop a severe aplastic anemia or pure red cell aplasia without evidence of exposure to a toxic drug or chemical. Many of these patients have evidence for T-cell suppression of erythroid progenitor cells and will respond to immunosuppressive therapies. In the case of pure red cell aplasia, the suppression of red cell growth is often related to the presence of a thymoma.

In children, red cell aplasia or aplastic anemia (pancytopenia) are seen as inherited defects. Constitutional aplastic anemia (Fanconi's anemia) is an autosomal recessive disorder where a severe aplastic anemia (pancytopenia) is associated with multiple physical defects. Blackfan-Diamond anemia is a congenital form of pure red blood cell aplasia. It is also associated with minor physical deformities and typically responds to low doses of prednisone.

The Maturation Disorders

The characteristic erythropoietic profile of the maturation disorders is summarized in Table 14. The combination of abnormal red cell morphology (macrocytosis or microcytosis) and a low reticulocyte index, together with a proliferative marrow (marrow E:G ratio greater than 1:1), are the key findings that suggest this group of disorders. The discrepancy between the reticulocyte production index and the number of marrow red cell precursors is typical of the ineffective erythropoiesis that accompanies most maturation disorders. Elevated bilirubin and LDH levels also demonstrate the increased cell turnover associated with ineffective erythropoiesis. Iron supply studies vary according to the individual disease state.

On the basis of red blood cell size and morphology, maturation disorder anemias are divided into two major categories: the *cytoplasmic maturation defects,* in which there is a decrease in hemoglobin synthesis resulting in microcytosis, and the *nuclear maturation defects,* characterized by megaloblastic changes within the marrow and macrocytosis. The prominence of these findings depends on the severity of the anemia. For example, early in the development of iron deficiency, patients present with a hypoproliferative anemia characterized by normocytic, normochromic red blood cell morphology. Microcytosis and hypochromia appear when the hemoglobin falls to levels below 10 g/dL and remains at that level for

Table 14 • **Maturation Disorders: Erythropoietic Profile**

Smear/Red Cell Indices
Microcytosis (MCV, <80 fL) or macrocytosis (MCV, >100 fL) with prominent anisocytosis and poikilocytosis

Reticulocyte Index
Less than 2.0
Polychromasia present

Marrow E:G Ratio
Greater than 1:1 when anemia is severe (Hb <10 g/dL)
Abnormal red cell precursor morphology (poor hemoglobinization with cytoplasmic maturation disorders; megaloblastosis with nuclear maturation disorders)

Indirect Bilirubin/LDH Levels
Increased (except with iron deficiency, in which bilirubin is decreased)

several months. The full picture of a cytoplasmic maturation defect with ineffective erythropoiesis appears only when the anemia is very severe, characterized by a hemoglobin below 5 to 6 g/dL. Similarly, a full-blown nuclear maturation defect secondary to vitamin B_{12} or folic-acid deficiency with macrocytosis and an ineffective, megaloblastic bone marrow takes time to evolve. The vitamin deficiency state must be sustained over a number of months, and the hemoglobin must fall below 8 to 10 g/dL.

The Cytoplasmic Maturation Defects

A defect in hemoglobin synthesis severe enough to result in a microcytic anemia can result from iron deficiency, a congenital defect in globin synthesis, or a disorder of mitochondrial function and porphyrin synthesis. Because iron deficiency is common in most populations, it is always a good strategy to evaluate iron supply as the first step in the workup of microcytosis. As described in the section on iron deficiency, iron deficiency anemia is readily diagnosed from the combination of low serum iron, increased TIBC, and very low serum ferritin levels (see Fig. 41). Moreover, the serum ferritin and TIBC levels permit separation of true iron deficiency from inflammation (see Table 10).

The Thalassemias

An inherited defect in globin-chain synthesis (thalassemia) is the other leading cause of a microcytic anemia.[61] The racial background of the patient often provides a clue to the specific defect involved. For some geographic populations, the incidence of thalasssemia exceeds that of iron deficiency, whereas in other areas, both are common and may present together in the same patient. Because of this, iron supply studies need to be performed in all patients, even when the presence of thalassemia appears certain.

The major categories of thalassemia and related hemoglobinopathies are summarized in Table 15. Clinically, the most important thalassemias are related to defects in the α- and β-globin genes. In the case of β thalassemia, one or both β-globin genes may be defective and, depending on the nature of the mutation, produce little or no change in hematologic status *(thalassemia minor* or *trait),* moderate anemia with well-defined microcytosis *(thalassemia intermedia),* or a severe form of the disease with marked anemia *(thalassemia major).* The variation in clinical presentation can be explained from the inheritance pattern of the β-gene defect, either

Table 15 • **Major Categories of Thalassemia**

β Thalassemia, heterozygote and homozygote forms
β° (deletion)
β^{+} (partial deficiency)
Hemoglobin Lepore
Hemoglobin E disease
β-δ Thalassemias
Hereditary persistence of fetal hemoglobin
α Thalassemia
$\alpha^0 \vert \alpha^0$ Hb Barts hydrops syndrome
$\alpha^0 \vert \alpha^0$
Compound α^0 α^+ heterozygotes, Hb H disease

complete deletion of one or both β genes (β°) or a partially defective gene that produces β chains at a reduced level (β^{+}). Actually, the genetic defects are extremely heterogeneous; more than 50 different mutations have been reported that affect the transcription, processing, and translation of mRNA. The severity of the clinical picture can also be affected by the coinheritance of an α-gene defect or other hemoglobinopathy. In Southeast Asia, hemoglobin E, a β-gene point mutation, is a common defect characterized by poor β-chain formation, either alone or as a double heterozygote with β thalassemia. α-Gene defects are also common in Asian and certain African populations. Because there are four α genes, the spectrum of α thalassemia can vary from no clinical disease to a lethal defect according to the number of genes involved.

Detection and accurate diagnosis of a thalassemia requires both clinical expertise and the skilled use of a number of laboratory tests. The first clue to the presence of a thalassemia is usually some combination of anemia, microcytosis, hypochromia, and targeting on a routine CBC. The clinical picture—that is, the severity of the anemia and microcytosis, family history, and racial background—may make the diagnosis easy (Fig. 44). For example, a severe microcytic, hypochromic anemia in a child whose parents are of Greek descent and known to have β thalassemia minor is almost certainly β thalassemia major. It can be useful, therefore, to organize the differential diagnosis around the clinical pattern of the anemia. However, proof of the exact nature of the thalassemia requires analysis of the patient's hemoglobin structure (Fig. 45).

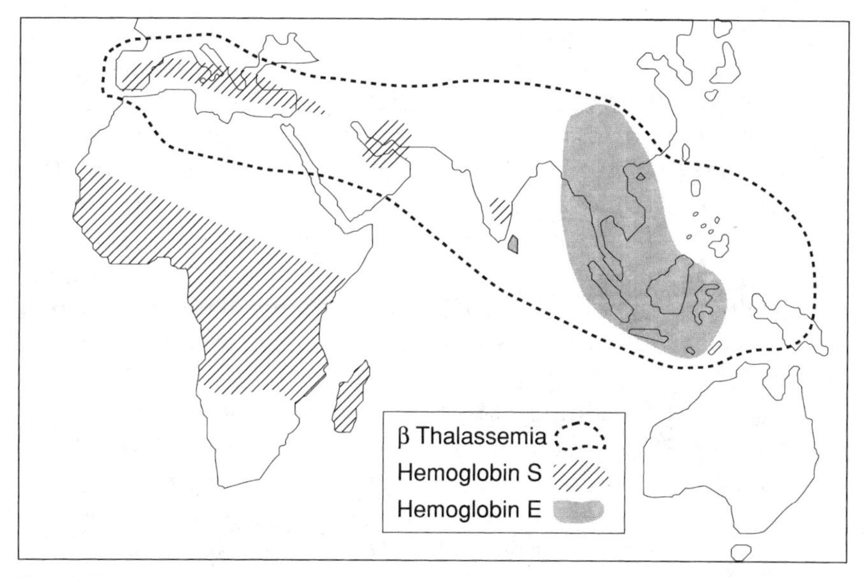

Figure 44. Geographic distribution of the major hemoglobinopathies. The broad distribution of β thalassemia is described by the dotted line. Hb S (sickle cell anemia) is most prevalent in central Africa but appears to a lesser extent in the Mediterranean basin, the Arabian peninsula, and southern India. Hb E is largely confined to Southeast Asia.

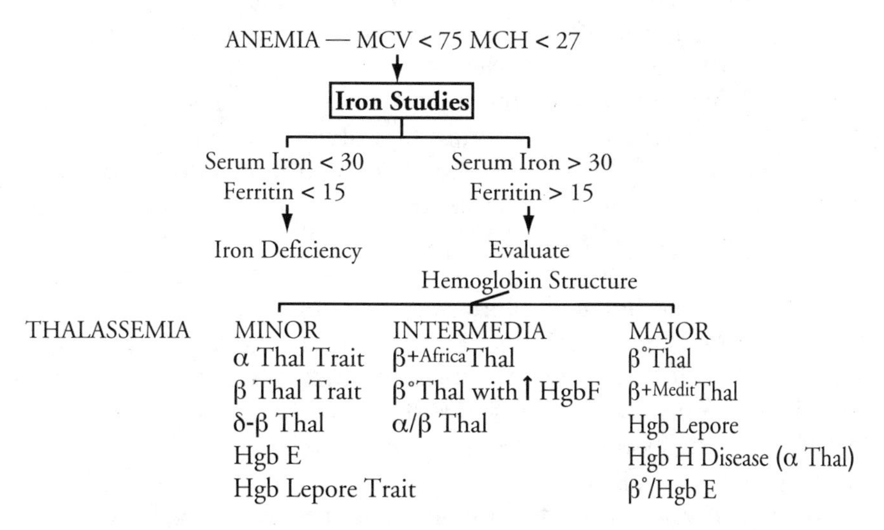

Figure 45. Approach to the diagnosis of iron deficiency and the more common thalassemias. Iron studies are used initially to separate severe iron deficiency from thalassemia. The differential diagnosis of the individual forms of thalassemia is organized around the clinical presentation and analyses of hemoglobin structure.

Patients with a single α-gene deletion or certain β-gene mutations are not detectable clinically; they are silent carriers of the gene defect. Thalassemia minor patients are defined by their microcytosis (a MCV less than 75 fL) and hypochromia, together with a mild or negligible anemia (hemoglobin of 10 to 14 g/dL). The most common thalassemia minor genotypes are a single β-gene deletion or mutation, β-δ thalassemia, or hemoglobin Lepore trait. α thalassemia minor patients have two α-gene deletions or mutations (called *hemoglobin Constant Spring*). Patients in the thalassemia intermedia category demonstrate a moderate anemia with marked microcytosis. The genotypes responsible for this more severe level of disease include homozygous β thalassemia of the $\beta^{+\text{Africa}}$ type, homozygous β° thalassemia with hereditary persistence of hemoglobin F (Hb F), and double heterozygotes for α and β thalassemia.

Very severe anemia with marked microcytosis is characteristic of patients with thalassemia major. The anemia presents early in childhood and generally requires chronic transfusion support. These children can have skeletal changes secondary to marrow expansion. They are short in stature and fail to mature sexually. Because of the severe ineffective erythropoiesis and recurrent transfusion, they experience progressive iron loading and can die from heart failure early in life. The genotypes responsible for thalassemia major include homozygous deletion or mutation of the β gene and the coinheritance of β thalassemia and hemoglobin E (Hb E). Patients with a deletion or mutation of three α genes present with microcytosis and a mild to moderate hemolytic anemia, rather than the ineffective erythropoiesis seen with β thalassemia. Involvement of all four α genes in a lethal defect, resulting in fetal loss late in pregnancy (hydrops fetalis).

The defects in the α and β genes that produce thalassemia can be inherited together with point mutations in the β gene (Hb E, Hb S, and Hb C). These patients demonstrate a mixed defect. Those who are double heterozygotes for β thalassemia and Hb E generally have the clinical features of β thalassemia major. The combination of β thalassemia and hemoglobin S results in a clinical picture similar to that of sickle cell anemia, although most patients demonstrate splenomegaly. In some patients, high levels of Hb F production protect against painful vaso-occlusive crises. Patients with β-δ thalassemia and Hb S or the combination of Hb S and hemoglobin Lepore trait have milder disease with few, if any, crises. Their anemia is much less severe, although microcytosis, hypochromia, targeting, and occasional sickle cells are apparent on the CBC. The decreased severity is also related to the production of higher levels of Hb F. Finally, the combination of Hb C and β thalassemia produces a moderately severe hemolytic anemia with microcytosis, hypochromia, and pronounced targeting.

Although the severity of the anemia, the red blood cell morphology, and the racial background of the patient may suggest a specific type of thalassemia, the definitive diagnosis depends on the analysis of the patient's hemoglobin pattern. Clinical laboratories can perform hemoglobin electrophoresis (cellulose acetate and citrate agar) to identify the most common hemoglobinopathies, including Hb S, Hb C, Hb E, Lepore trait, and Constant Spring. The laboratory can also quantitate the amount of Hb A_2 and Hb F to diagnose β thalassemia. The distribution of Hb F in red cells and the presence of Hb H as a marker for α thalassemia can be evaluated using the supravital stain, brilliant cresyl blue, while quantitation of hemoglobin H is possible using starch gel electrophoresis. The patterns of abnormal hemoglobins for the most common thalassemias are summarized in Table 16.

Patients with thalassemia major or intermedia should also have their iron status evaluated. Marked ineffective erythropoiesis is associated with increased iron absorption and a rise in the plasma

Table 16 • **Hemoglobin Patterns in Thalassemia**

% Hemoglobin	A	F	A_2	E or S	H (β_4)
Normal	97	<1	2–3		
β Thalassemia minor	80–95	1–5	3–7*		
β Thalassemia intermedia	30–50	50–70	0–5		
β Thalassemia major	0–20	80–100	0–5		
α Thalassemia trait	85–95				Red cell inclusions
Hgb H disease	60–95				5–30
Hgb E trait	60–65	1–2	2–3	30–35	
Hgb E homozygous	0		5	95	
E/β Thalassemia	0	45	1–5	40–50	
S/β Thalassemia	0–30	5–15	—	50–90	

*Iron deficiency lowers the hemoglobin A_2 level and conceals the diagnosis of β thalassemia minor.

iron level to full saturation of transferrin. This results in iron loading of hepatic parenchymal and cardiac muscle cells. Transfusion makes matters worse by adding to the iron burden. All patients should, therefore, have periodic measurements of their plasma iron, TIBC, and ferritin levels.

The prevention of the more severe forms of thalassemia is now possible with screening of high-risk populations and prenatal diagnosis. At-risk mothers and fathers should be tested early during a pregnancy to identify possible hemoglobinopathy combinations. In utero diagnosis of fetal hemoglobinopathies is possible by amniocentesis or chorionic villus sampling for fetal DNA analysis. Gene deletions can be diagnosed directly by Southern blot, and nondeletional thalassemias can be detected using restriction (fragment) length polymorphism studies of both parental and fetal tissue.

The Sideroblastic Anemias

Disorders of mitochondrial function and porphyrin synthesis can also produce a cytoplasmic maturation defect. The hallmark of these conditions is the appearance of ring sideroblasts on the iron stain marrow preparation. By light microscopy, the typical ring sideroblast has a partial or complete ring of relatively uniform iron granules encircling the nucleus (see Fig. 32). On electron microscopy studies, the iron is found to be deposited between the cristae of morphologically distorted mitochondria. True ring sideroblasts should not be confused with normoblasts containing excess cytoplasmic iron. The latter result when iron supply is increased and globin synthesis is decreased (the thalassemias) and in alcoholic patients who have an increased plasma iron level, a folate-deficient megaloblastic anemia, and alcohol damage to the mitochondria.

Defects in mitochondrial function, clinically referred to as the sideroblastic anemias, represent a heterogeneous group of conditions (Table 17).[62] Hereditary forms of sideroblastic anemia are transmitted as an autosomal or sex-linked trait with variable penetration and expression. The severity of the anemia differs from individual to individual, leading to the detection of the disease during childhood in some patients but not in others. In the adult, most sideroblastic anemias are acquired and are found in association with a variety of drugs, neoplasms, and inflammatory conditions. When none of these conditions is present, and in the absence of familial involvement, it is assumed that the patient has a primary

Table 17 • **Sideroblastic Anemias**

Hereditary
Sex-linked
Autosomal dominant or recessive
Congenital—inheritance not determined
Acquired
Idiopathic
Myelodysplastic (RARS)
Drug-related (isoniazid, chloramphenicol)
Chemical-related (lead, alcohol, zinc toxicity)
Associated with other disorders (neoplasias, inflammatory states)

marrow disorder referred to as acquired, idiopathic sideroblastic anemia or refractory anemia with ring sideroblasts (RARS).

The erythropoietic profile of the sideroblastic anemias ranges from the typical picture of a cytoplasmic maturation defect to a confusing combination of both nuclear and cytoplasmic maturation abnormalities.[63] Hereditary sideroblastic anemia is typically microcytic and hypochromic, although there is usually a double population of microcytic and normocytic cells. With the acquired sideroblastic anemias, the MCV may be either decreased, normal, or increased. The blood smear, showing microcytic, hypochromic cells, may also show a few macrocytes. The degree of anisocytosis and poikilocytosis is proportionate to the degree of ineffective erythropoiesis. Some RARS patients have very ineffective erythropoiesis; others have a hypoproliferative anemia with few changes in blood and marrow morphology.

With time, granulocytopenia and thrombocytopenia are frequent findings in RARS patients, compatible with a pluripotent stem-cell defect. Most of these patients have an underlying clonal malignancy and therefore have been classified as having one of a group of myelodysplastic syndromes that includes refractory anemia (RA), refractory anemia with ring sideroblasts (RARS), refractory anemia with excess blasts (RAEB), refractory anemia with excess blasts in transformation (RAEB-t), and chronic myelomonocytic leukemia (CMML). The overall prognosis depends on the rate of appearance of granulocytopenia and thrombocytopenia and the appearance of abnormalities in the chromosomal karyotype. Patients with marked ineffective erythropoiesis are vulnerable to iron overload from transfusion and excessive iron absorption, whereas patients with hypoplastic marrows usually progress to either marrow failure or leukemia.

The Nuclear Maturation Defects

An increase in the MCV or the appearance of macrocytic red blood cells on the peripheral smear strongly suggests a nuclear maturation defect. However, other causes of macrocytosis must first be excluded (Table 18). The most common explanations for a moderate increase in the MCV without an apparent nuclear maturation defect are stimulated erythropoiesis, alcoholism, and liver disease. High levels of erythropoietin stimulation cause a shift of marrow reticulocytes into circulation. These cells can have an MCV in excess of 140 fL. In patients with hemolytic anemias, the uncorrected reticulocyte count can increase to levels in excess of 20% and, because of this, increase the MCV to 120 to 130 fL.

More moderate macrocytosis (MCV of 100 to 115 fL) is seen in patients with liver disease. In this case, the red blood cells are of uniform size but appear relatively thin with target cell formation. If there is superimposed hemolysis secondary to hypersplenism, the added reticulocytosis can further increase the MCV. "True" macrocytosis associated with a well-developed nuclear maturation defect can be identified by its characteristic morphology. The macrocytes seen in patients with vitamin B_{12} or folic acid deficiency often have an oval shape and are referred to as macro-ovalocytes. Their volume is up to twice that of other red blood cells because they result from a skipped mitosis. They are also well filled with hemoglobin, and on smear appear thick and hyperchromic, without the basophilia of the shift reticulocyte.

Table 18 • **Spectrum of Macrocytic Indices**

	Normal	Stimulated Erythropoiesis	Liver Disease	Nuclear Defect
MCV (fL)	90+8	110+20	105+10	120+30
Fragmentation (poikilocytosis)	0	0	0	+
Red cell morphology	Normal	Shift macrocytes	Targets	Macro-ovalocytes

The presence of even a small number of macro-ovalocytes on the peripheral blood smear can be an early clue to the presence of a nuclear maturation defect secondary to either vitamin B_{12} or folic-acid deficiency. They may be present even before anemia has developed or the MCV has increased. The presence of granulocytes with five or six lobes (hypersegmented polymorphonuclear leukocytes) is also a clue to a macrocytic anemia. However, uniform hypersegmentation is seen as an inherited trait without implications as to the maturation of other hematopoietic elements. Therefore, as a solitary finding, it is not diagnostic.

The characteristic erythropoietic profile of a nuclear maturation disorder is summarized in Table 19. The combination of macrocytosis, a low reticulocyte index together with a proliferative marrow (marrow E:G ratio greater than 1:1), and megaloblastic morphology are the key findings that suggest a nuclear maturation defect, especially one secondary to vitamin B_{12} and folic acid deficiency. In patients who are severely anemic, with marked ineffective erythropoiesis, both bilirubin and LDH levels are increased, even to the point where the patient appears to be jaundiced. Iron supply studies in these patients typically show high serum iron, normal TIBC, and normal to elevated serum ferritin levels. The elevation in the serum iron level is again a reflection of the ineffective erythropoiesis.

The differential diagnosis of a nuclear maturation defect involves separating the two vitamin deficiency states (vitamin B_{12} and folic acid deficiency) from other abnormalities in DNA syn-

Table 19 • **Nuclear Maturation Disorder: Erythropoietic Profile**

Smear/Red Cell Indices
- True macro-ovalocytes with prominent anisocytosis and poikilocytosis
- MCV = 100–140 fL

Reticulocyte Index
- Less than 1.0
- Polychromasia present

Marrow E:G Ratio
- Greater than 1:1
- Erythroid hyperplasia with megaloblastic morphology

Indirect Bilirubin/LDH Levels
- Increased because of ineffective erythropoiesis

thesis, including the predictable effects of chemotherapeutic drugs, the rare hereditary abnormality in nucleic acid metabolism, and acquired, intrinsic abnormalities in DNA synthesis. An abnormality secondary to a chemotherapeutic agent is generally obvious based on the patient's history and its reversibility once the drug is withheld. If this is not the case, the evaluation should begin with studies of the patient's vitamin B_{12} and folic acid status because this leads directly to effective therapy. Other acquired, intrinsic abnormalities of nuclear maturation can be associated with preleukemic change and are less amenable to therapy.

Vitamin B_{12} Deficiency

Vitamin B_{12} deficiency is almost always associated with impaired absorption of the vitamin. Vitamin B_{12} is ubiquitous in diets containing animal byproducts, and normally 3 to 5 μg of the vitamin are absorbed each day. Vegetables do not contain vitamin B_{12} unless they are contaminated with microorganisms capable of synthesizing the vitamin. Therefore, true vegetarians who do not fortify their diets are at risk of developing a deficiency state.

The absorption of vitamin B_{12} is a complex process (Fig. 46). Vitamin B_{12} in food is initially bound to a salivary binding protein (R-protein) until it reaches the upper portion of the small bowel, where it is liberated by pancreatic proteases for binding to the glycoprotein, intrinsic factor. This complex passes through the small intestine until it reaches and binds to a receptor on the mucosal cells of the ileum. The vitamin is then released from intrinsic factor and transported across the gut wall to form a complex with transcobalamin II (TC II), a plasma β globulin, for transport to tissues. Although some B_{12} is directly transported to the erythroid marrow, a portion is also stored in liver parenchymal cells. Under normal circumstances, 1 to 10 mg of vitamin B_{12} are stored in the liver. This is sufficient to support erythropoiesis for 2 to 5 years if vitamin B_{12} absorption is suddenly disrupted. However, the actual turnover rate of the liver B_{12} stores depends on their size and tissue metabolic needs. It can range from as little as 0.5 μg to as much as 8 μg per day. Several micrograms of the vitamin are secreted into bile each day for reabsorption. In the presence of small-bowel disease, especially ileal disease, interruption of this enterohepatic cycle can increase the rate of vitamin loss in the stool.

Vitamin B_{12} deficiency in adults is almost always caused by a defect in absorption (Table 20).[64] Autoimmune disease with anti-

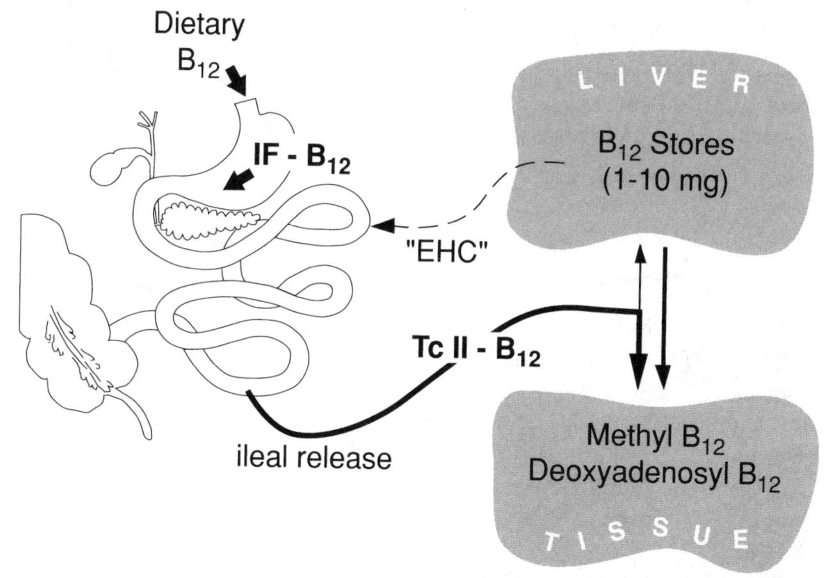

Figure 46. The absorption, transport, and storage of vitamin B_{12}. Dietary vitamin B_{12} is liberated by peptic acid digestion and then bound to intrinsic factor (IF-B_{12}) produced by the gastric parietal cells. This protects the vitamin B_{12} until it reaches the ileum, where it is released from the intrinsic factor and bound to transcobalamin II (TC II) for transport to liver and tissues. In normal individuals, 1 to 10 mg of vitamin B_{12} are stored in the liver. These B_{12} stores may be released and transported directly to tissues as methyladenosyl B_{12} or deoxyadenosyl B_{12}, or to a lesser extent secreted into bile for reabsorption, an enterohepatic cycle (EHC).

body specific for parietal cells, intrinsic factor, or the intrinsic factor–vitamin B_{12} complex (IF-B_{12}) is implicated in many patients and may have a family prevalence. Severe gastric atrophy or gastric surgery can also result in a loss of intrinsic factor secretion. Pancreatic insufficiency may interfere with the release of vitamin B_{12} from R-protein, preventing binding to intrinsic factor. Certain intestinal parasites and overgrowth of bacteria in the small intestine consume the dietary vitamin B_{12} before it reaches the ileum. Any intrinsic disease of the ileal mucosal cells or surgical resection of the ileal portion of the small intestine will markedly affect absorption. Patients with acquired immunodeficiency syndrome (AIDS) demonstrate a malabsorption of vitamin B_{12}, which can contribute to the severity of their anemia.

In addition to these absorption defects, a congenital deficiency of the plasma protein TC II can result in an inadequate supply of

Table 20 • **Causes of Vitamin B_{12} Deficiency**

Autoimmune Disease
- Anti-intrinsic factor, antiparietal cell antibodies

Intrinsic Factor Deficiency
- Congenital
- Atrophic gastritis
- Gastrectomy

Malabsorption
- Pancreatic insufficiency
- Zollinger-Ellison syndrome
- Bacterial overgrowth
- Intestinal parasites
- Sprue
- Small-bowel lymphoma
- Regional ileitis or ileal resection
- AIDS

Defective Transport/Storage
- Transcobalamin II deficiency

Metabolic Defect
- Nitrous oxide anesthesia
- Congenital abnormalities

vitamin B_{12} to tissues despite a normal diet and intestinal absorption. This may be associated with a normal plasma B_{12} level because the amount of the vitamin B_{12} bound to the other two transport proteins, transcobalamin I and III (TC I and TC III) is unaffected. A vitamin B_{12} deficiency state can be induced by nitrous oxide anesthesia secondary to its effect on the stability of the B_{12} molecule. Finally, intracellular vitamin B_{12} metabolic defects have been reported in children with methylmalonic aciduria and homocystinuria.

Vitamin B_{12} deficiency in adults was originally called *pernicious anemia* inasmuch as it was uniformly fatal before vitamin B_{12} therapy. Because the lack of the vitamin affects both the hematopoietic and nervous systems, the term *combined systems disease* was also popularized. The sensitivity of the hematopoietic system to B_{12} deficiency reflects the high turnover of marrow precursors and their requirement for vitamin B_{12} to support DNA synthesis. Neurologically, vitamin B_{12} deficiency causes a distinctive demyelinating lesion of the peripheral nerves, posterior lateral columns of the spinal column, and occasionally of the brain. Vibratory and position sense are impaired at an early stage. In addition to a peripheral neuritis manifest by paresthesias of the hands and feet and diminution of sensation, patients complain of mood changes, loss of memory, and confusion. Usually the macrocytic anemia is present when the patient develops neurologic symptoms and signs, making the diagnosis easy. However, in an occasional patient, a progressive neuropathy may be observed without the accompanying megaloblastic anemia.

Folic Acid Deficiency

Folic acid deficiency produces a megaloblastic anemia that is indistinguishable from pernicious anemia, though without the neuropathy of B_{12} deficiency. It is usually associated with a deficient dietary intake of folate in the face of an increased body requirement. The minimum daily requirement for folic acid in a normal adult is approximately 50 μg. However, this requirement is increased two- to fourfold during pregnancy and in patients with high rates of cell turnover, such as those with hemolytic anemia. On the supply side, standard Western diets can contain much less than 500 μg of absorbable folate, setting the stage for a dietary deficiency state. This vulnerability is especially pronounced in chronically alcoholic patients, whose dietary intake of folate is minimal.[65]

The pathways of absorption and distribution of folate are shown in Figure 47. Dietary folate is absorbed in the proximal portion of the small intestine, where mucosal cells are rich in dihydrofolate reductase. It enters the plasma as methyltetrahydrofolate and is directly transported to tissues, where it serves intracellular metabolic pathways. Certain plasma proteins bind and transport folate, especially the nonmethylated analogues, which are recycled to the liver for remethylation and storage.

Liver parenchymal cells accumulate folate stores in the form of methyltetrahydrofolate polyglutamates. These stores are in dynamic equilibrium with tissues through the folate enterohepatic cycle. Faced with an intermittent dietary supply of folate, the enterohepatic cycle serves to maintain a constant delivery of folate to the gut for absorption and transport to tissues, providing up to 200 μg or more of folate each day.[66] Diseases of the small intestine, especially tropical and nontropical sprue, are associated with the rapid onset of a severe folate-deficient megaloblastic anemia. In this situation, the defect in intestinal absorption not only prevents

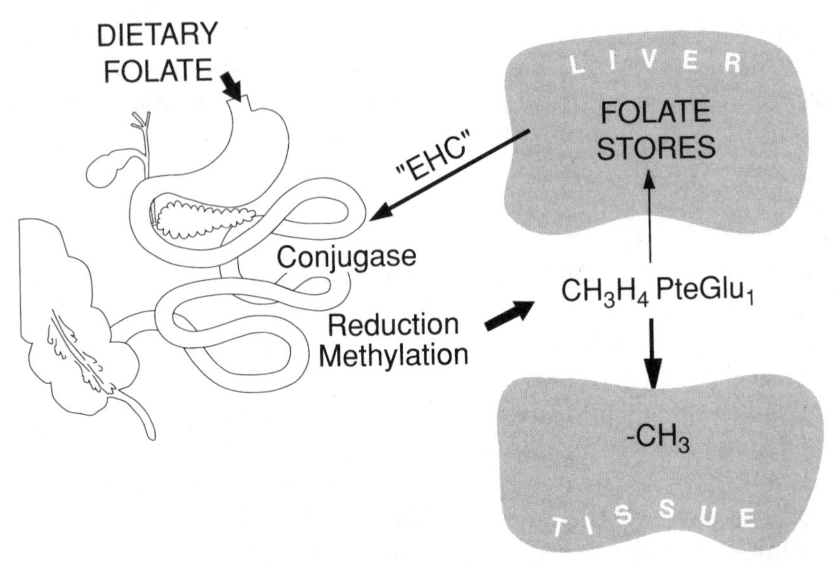

Figure 47. Absorption, transport, and storage of folic acid. Dietary folate polyglutamate is hydrolyzed (conjugase), reduced, and methylated to methyltetrahydrofolate monoglutamate before transport to tissues and liver. Most of the folate absorbed goes directly to tissue to act as a methyl donor. A smaller amount is stored in the liver as methyltetrahydrofolate polyglutamate. A constant turnover of liver folate stores and secretion into bile forms an important enterohepatic cycle (EHC), which helps maintain a constant supply of folate to the intestine, and subsequently to body tissue.

the adequate uptake of folate from the diet but also interferes with the enterohepatic cycle. Alcohol also disrupts the recycling of folate from liver stores to tissues, thereby restricting folate supply.[67] Although a normal individual can sustain normoblastic erythropoiesis for several months on a folate-deficient diet, a macrocytic, megaloblastic anemia can appear in alcoholic subjects within a few weeks of sustained alcohol intake.[68]

Diagnosis

The diagnosis of a vitamin B_{12} or folate deficiency anemia may be possible from the clinical presentation. Folate deficiency is commonly seen in chronically alcoholic patients, especially those who are eating little or no food. Pregnant women and patients with hemolytic anemias are also at risk. In contrast, vitamin B_{12} deficiency is most often associated with diseases of the stomach or ileum; only strict vegetarians are at risk for a dietary deficiency state. Another obvious clue to a vitamin B_{12} deficiency is the simultaneous presence of a macrocytic, megaloblastic anemia and the neurologic symptoms typically seen with B_{12} deficiency. Folate deficiency is rarely, if ever, associated with abnormalities in vibratory and position sense, although sensory neuropathies are common in chronically alcoholic patients.

The erythropoietic profiles of the two deficiency states are virtually identical (see Table 19). In gastrointestinal disease (sprue, short bowel syndrome, and so on), multiple deficiency states can be present in the same patient because of simultaneous malabsorption of folic acid, vitamin B_{12}, and iron. Gastrointestinal bleeding can also contribute to the final anemia profile. For example, blood loss and iron deficiency can produce a dimorphic anemia in which the MCV is never a normal but red cells have a decreased hemoglobin content. Iron deficiency also tends to suppress the level of marrow proliferation and the severity of the megaloblastic morphology. In other patients, the megaloblastic erythropoiesis and macrocytosis can overshadow the iron deficiency state. The iron deficiency component only emerges when the patient is treated with the appropriate vitamin, either vitamin B_{12} or folic acid.

The erythropoietic profile of a macrocytic anemia in an alcoholic patient can be confused by the timing of the patient's workup. The full-blown picture of a macrocytic, megaloblastic anemia with low folate levels is characteristic of the chronically alcoholic patient who maintains blood alcohol levels above 100 mg/dL for extended periods. As soon as alcohol intake is interrupted, the enterohepatic cycle and the absorption of folate from the diet quickly return to normal, resulting in a rapid conversion of the marrow to normoblastic morphology. Within a week or two, the patient even demonstrates a brisk reticulocytosis, suggesting the presence of a hemorrhagic or hemolytic anemia, not the original megaloblastic anemia. It is important, therefore, to evaluate a macrocytic anemia as soon as it is first detected and before the patient returns to a normal diet or is treated with folic acid, vitamin B_{12}, or iron.

A number of laboratory studies are available to help in the diagnosis of a macrocytic anemia (Table 21).[69] Measurements of serum vitamin B_{12} and folic acid levels are of the greatest value in detecting a shortage of either vitamin. They are also the two tests that change most quickly once therapy is initiated. Measurements of serum or urine methylmalonic acid and homocysteine levels are very sensitive detectors of a tissue deficiency state and have been

Table 21 • **Tests of Vitamin Deficiency**

	Normal	Possible Deficiency	Deficiency
Serum B_{12} level (pg/mL)	>200	100–200	<100
Serum folate (pg/mL)	>4	3–4	<3
Red cell folate (pg/mL)	>200	100–200	<100
Urinary methylmalonate (mg/24 h)	<4	4–10	>10
Gastric acidity (pH after histamine)	<3.5	—	>3.5
Schilling test (% excretion)	>7	2–7	<2

used to screen populations for subclinical vitamin B_{12} deficiency. However, these measurements also need to be done before any therapy. A deficiency of vitamin B_{12} should be suspected whenever the serum B_{12} level is less than 200 pg/mL, although there is considerable overlap with normal, and B_{12} levels between 100 and 200 pg/mL may not be diagnostic of a deficiency state. If the patient is megaloblastic and responds to specific vitamin therapy, a true tissue vitamin deficiency is likely. Otherwise, proof of a deficiency requires an assessment of cellular B_{12} metabolism with a test such as the urinary methylmalonic acid assay. Another test that can be used for this purpose is the deoxyuridine suppression test, which directly measures the response of lymphocytes to each vitamin.

Whenever vitamin B_{12} deficiency is detected, it is important to identify its cause. In the adult patient, the most cost-effective first step is to screen for serum autoantibodies against intrinsic factor, the cobalamin-intrinsic factor complex, and parietal cells. More than 60% of adult patients with pernicious anemia demonstrate one or another of these antibodies, suggesting an autoimmune basis for their disease. A less specific measurement is the determination of gastric acidity because acid and intrinsic factor secretion are lost simultaneously with atrophy or destruction of gastric parietal cells. To identify a defect in intestinal absorption, a Schilling test can be performed, first without and then with added intrinsic factor. This test involves the oral administration of a small amount of radiolabelled vitamin B_{12} while the patient is fasting and the subsequent measurement of the appearance of that radioactivity in serum and urine. Normal subjects excrete more than 7% of the radioactive dose in the first 24 hours after a flushing dose of nonradioactive B_{12}. Patients with gastric atrophy and intrinsic factor deficiency excrete less than 7% but demonstrate increased absorption and urinary excretion when intrinsic factor is added. In contrast, patients with diseases of the small intestine, such as sprue, bacterial overgrowth, parasitic infestation, small-bowel tumors, or resections, do not correct their vitamin B_{12} malabsorption.

Measurements of the serum vitamin B_{12}-binding proteins, TC I, TC II, and TC III, and the amount of vitamin B_{12} bound to each protein can provide a more sensitive measurement of vitamin B_{12} supply to tissues. Most of the B_{12} needed by bone marrow cells is transported by TC II, and the turnover of the cobalamin–TC II complex in serum is much more rapid than that of the other two binding proteins. A direct measurement of the serum level of cobalamin–TC II complex, like the measurement of the serum iron in an iron deficiency, can detect an early vitamin B_{12} deficiency state, even before the evolution of a megaloblastic anemia. Normal individuals have cobalamin-TC II levels greater than 50 pg/mL; negative balance is indicated when values fall below 40 pg/mL. This is a difficult assay, however, and is not offered by most laboratories.

Macrocytic anemias are also seen in patients with acquired abnormalities of DNA synthesis and certain inborn errors of nucleic acid metabolism. Patients who receive certain chemotherapeutic agents can develop a macrocytic anemia as well as leukopenia and thrombocytopenia. Antimetabolite drugs that interfere with the synthesis of DNA, RNA, and specific proteins, including the inhibitors of purine and pyrimidine synthesis (6-mercaptopurine, 6-thioguanine, azathioprine, and 5-fluorouracil), drugs that affect deoxyribonucleotide synthesis (hydroxyurea and cytosine arabinoside), and the dihydrofolate reductase inhibitors (aminopterin, methotrexate, triamterene, and trimethoprim) can all cause a nuclear maturation defect with a megaloblastic, macrocytic anemia. The inhibition of dihydrofolate reductase by folate antagonists can be reversed by the administration of formyltetrahydrofolate (citrovorum factor, leucovorin). The nuclear maturation defects associated with other chemotherapeutic agents resolve spontaneously when the drugs are withdrawn.

When a macrocytic anemia with megaloblastic marrow morphology is detected in childhood, a hereditary abnormality affecting either the B_{12} transport proteins (TC II deficiency) or nucleic acid metabolism should be suspected.[70] A number of congenital defects of folate metabolism have also been reported, including deficiencies of one or more enzymes in the folate metabolic pathways. In Lesch-Nyhan syndrome, a deficiency of the enzyme hypoxanthine guanine phosphoribosyltransferase, which is required in purine metabolism, is responsible. Hereditary orotic aciduria, a disorder of pyrimidine metabolism, is associated with a megaloblastic anemia that is unresponsive to vitamin B_{12} and folate therapy.

When a macrocytic anemia in an adult is unresponsive to vitamin B_{12} or folic acid, it is generally a sign of a premalignant condition. Such patients often show marrow hyperplasia with varying degrees of megaloblastosis. Frequently, there is a marked dyspoiesis with striking abnormalities in nuclear development involving late erythroid precursors. Involvement of the granulocytic and megakaryocytic maturation sequence is also seen, although the hypersegmentation of circulating granulocytes that is so typical of vitamin B_{12} and folic acid deficiencies is rarely present. Many patients demonstrate a simultaneous impairment in hemoglobin synthesis with the appearance of ring sideroblasts in the marrow. The same patients may have the distinctive finding of periodic acid-Schiff (PAS)-positive granules in the cytoplasm of their marrow normoblasts. These patients have a chronic, slowly progressive course and develop either leukemia or a severe pancytopenia with a hypoplastic bone marrow. The latter can resemble either acute myelocytic leukemia or erythroleukemia.

Within the category of refractory macrocytic anemias with ineffective erythropoiesis is a rare group of hereditary refractory anemias termed *congenital dyserythropoietic anemia (CDA).* Three subclasses of this disorder have been described, all characterized by a striking erythroid precursor multinuclearity. CDA type I presents with slight to moderate macrocytosis and marked marrow erythroid hyperplasia with some megaloblastic features. The diagnostic morphologic feature is the appearance in the marrow of small numbers of cells with double nuclei. Type II CDA, also called *hereditary erythroblastic multinuclearity with a positive acid serum test (HEMPAS),* is more common. In this condition, the erythroid precursors in the marrow show marked multinuclearity, and circulating red cells show a susceptibility to lysis by an acid pH and by certain ABO-compatible sera. Finally, CDA type III is characterized by giant macrocytes on smear and extremely large marrow megaloblasts, each with as many as 12 nuclei.

The Hemorrhagic/Hemolytic Anemias

As a class, the hemorrhagic and hemolytic anemias are distinguished by their normocytic, normochromic red cell morphology and a pronounced increase in red cell production levels to greater than three times basal levels. However, the erythropoietic profile following acute hemorrhage or hemolysis depends on the time course of the anemia. The full erythroid marrow response with a rise in the reticulocyte index requires a steady state anemia of over a week's duration and an adequate nutrient supply to the marrow. This is especially true in blood loss, in which an adequate supply of iron stores is essential.

The distinction between hemorrhagic and hemolytic anemia may be obvious from the clinical presentation. Massive bleeding from either the upper or lower gastrointestinal tract is associated with hematemesis and/or black or bloody stools. Massive external bleeding from trauma is even more obvious. Internal bleeding can be more difficult to detect and therefore is not as easily distinguished from a hemolytic process. When red blood cells are extravasated into tissues, they are gradually catabolized by scavenger cells, resulting in a process similar to that of a hemolytic anemia. The only difference is the accompanying signs of blood volume depletion, which can occur with severe bleeding, and the lower rate of catabolism of extravasated red blood cells, which can limit the erythropoietic response. Patients with acute hemolysis, resulting in a moderate to severe anemia, typically show a rapid, pronounced reticulocytosis.

Acute Blood Loss Anemia

Rapid blood loss amounting to more than 20% of the circulating blood volume produces symptoms varying from mild orthostatic hypotension and tachycardia to syncope and shock (Table 22). This is purely a reflection of blood volume losses. The anemia that results is at first hidden and appears only after sufficient time has passed for the plasma volume to expand and reconstitute the total blood volume (Fig. 48). Without volume replacement therapy, this can take 1 to 3 days. If effective fluid therapy is administered, the hemoglobin (hematocrit) will fall rapidly as the plasma volume expands.

A normal individual with a normal marrow and good iron stores is well equipped to respond to acute blood loss with an increase in red blood cell production. As a rule of thumb, once the

Table 22 • **Signs and Symptoms of Acute Blood Loss**

Less than 1000 mL or 20% TBV	No signs or symptoms, vasovagal reaction on occasion
More than 1000 mL or 20% TBV	Orthostatic hypotension; tachycardia with exercise
More than 1500 mL or 30% TBV	Fall in supine blood pressure; rest tachycardia; anxiety or restlessness
More than 2000 mL or 40% TBV	Hypovolemic shock, air hunger, disorientation, diaphoresis

hemoglobin level falls below 10 g/dL, the normal individual responds with a reticulocyte index of at least three times the basal level.[36] How long this is sustained depends on the availability of iron stores or an iron supply from other sources. Normal hepatocyte and reticuloendothelial iron stores can support production levels of three times basal levels for a relatively short period in patients experiencing ongoing blood loss. In contrast, patients with hemorrhage into tissues do much better because they are able to

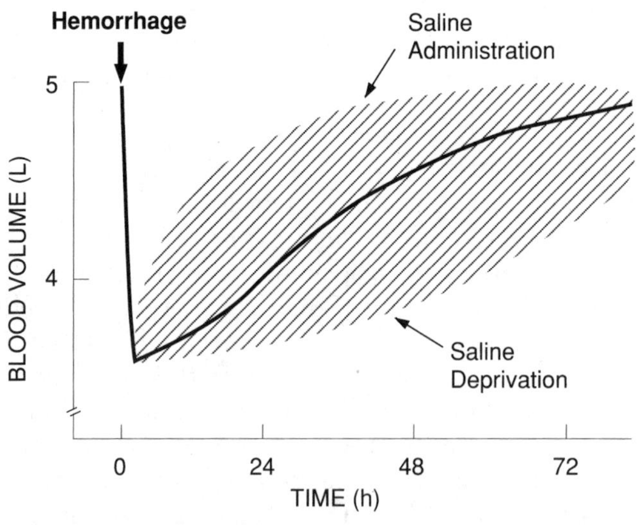

Figure 48. Changes in blood volume following hemorrhage. Restoration of the blood volume following acute bleeding requires an expansion of the plasma volume, which in the absence of volume replacement therapy can take 1 to 3 days. Saline administration speeds the process, whereas saline deprivation retards it.

harvest the iron from the breakdown of the extravasated red blood cells. Patients with external blood loss who receive continuous oral iron supplementation can also maintain red cell production levels of two to three times normal.

It is also important to recognize that there is a time lag before red blood cell production shows a maximal response to acute blood loss. As shown in Table 23, the full response of the marrow does not occur for 7 to 10 days after a hemorrhagic event. During the first few hours of the appearance of a blood-loss anemia, there is a rapid release of erythropoietin, which does produce a shift of marrow reticulocytes into circulation. Although this results in an increase in the reticulocyte percentage, the production index remains at or about the basal level. Over the next several days, a wave of proliferation of marrow erythroid precursors is observed, beginning with immature basophilic normoblasts and progressing with time to polychromatic and orthochromatic normoblasts. A true rise in the reticulocyte production index occurs only when this maturation sequence is complete. Under the best of circumstances, this takes 4 to 7 days, with a maximum response at 7 to 10 days. If these normal characteristics of the bone marrow response are not taken into account, an evaluation soon after acute bleeding may be interpreted as either a hypoproliferative or ineffective marrow response, leading the clinician in the wrong direction.

Some patients are unable to significantly increase erythropoiesis following hemorrhage. Chronic blood loss predictably leads to exhaustion of iron stores. Iron depletion then limits red blood cell production; the E:G ratio falls below 1:2 , and the reticulocyte production index falls below twice the basal level. Initially, circulating red blood cells continue to be normocytic and normochromic

Table 23 • **Response to Acute Bleeding**

Time after onset (d)	0–2	3–6	7–10
Polychromasia	+	++	+++
Reticulocyte index	1	1–2	3–4
E:G ratio	1:3	2:3–1:1	1:1
Erythropoietic pattern	Hypoproliferative	Proliferation appears ineffective	Proliferation appears effective

with little anisocytosis or poikilocytosis. However, if blood loss continues without iron replacement, the resulting negative iron balance produces a more obvious impairment of hemoglobin synthesis. Microcytic, hypochromic red blood cells appear in circulation with an increase in both anisocytosis and poikilocytosis. This first produces a mixed population of microcytic and normocytic cells. Because a full turnover of circulating red blood cells takes up to 100 days, a uniform population of microcytes with a major fall in the MCV can take a month or more to appear after iron stores are exhausted. An increase in the RDW, reflecting the size variation of a mixed population of red blood cells, can occur much earlier.

Acute Hemolysis

The clinical presentation of a patient with acute hemolysis depends on the rapidity of the event, the amount of red cell destruction, and whether the damage produces significant hemoglobinemia and hemoglobinuria. To guide diagnosis, acute hemolysis is divided into two categories: *intravascular destruction,* in which there is appreciable lysis of red cells in circulation, and *extravascular destruction,* in which red blood cells are destroyed by phagocytosis. This distinction is somewhat artificial because overlap occurs. However, the presence of hemoglobinemia and hemoglobinuria is an important clue to certain types of intravascular hemolysis (Table 24). Massive intravascular hemolysis is important not only as a cause of severe anemia, but also because it can result in renal damage and even anuria.

Table 24 • **Causes of Intravascular Hemolysis**

Mechanical
- March hemoglobinuria, prosthetic valve fragmentation, vasculitis, consumptive coagulopathy

Osmotic
- Distilled water, embolus

Chemical or Thermal Damage
- Alpha toxin of *C. perfringens,* snake venoms, burns, drug and fava bean lysis with G6PD deficiency

Complement Damage
- Isoantibody plus complement, cold agglutinins, paroxysmal nocturnal hemoglobinuria

Acute intravascular hemolysis may be anticipated in certain clinical situations, such as severe thermal burns, malaria, certain snake bites, mixed anaerobe or *Clostridium perfringens* infection resulting in tissue necrosis, or transfusion with ABO-incompatible blood. Intravascular hemolysis can also occur following the ingestion of fava beans in patients with glucose 6-phosphate dehydrogenase (G6PD) deficiency and in mycoplasma pneumonia secondary to the appearance of a cold agglutinin. The laboratory diagnosis of intravascular hemolysis depends on the severity of the event and the timing of the evaluation. Even a small amount of free hemoglobin (the amount of hemoglobin released by the lysis of as little as 10 mL of red blood cells) tints the plasma pink for a few hours, with a plasma hemoglobin level of greater than 50 mg/dL. This small amount of hemoglobin is bound to the serum protein haptoglobin, which prevents its escape into the urine. However, the hemoglobin-haptoglobin complex is cleared by the liver parenchymal and reticuloendothelial cells, resulting in a measurable fall in the serum haptoglobin level.[71] When enough hemoglobin is released to exceed the haptoglobin-binding capacity, free hemoglobin passes through the glomerulus into the urine. A random urine sample shows either the red of hemoglobin or the brown of methemoglobin.

The time lapse from the initial hemolytic event is critical in interpreting laboratory findings of intravascular hemolysis (Fig. 49). If the patient is evaluated immediately after an acute, self-limited intravascular hemolytic event, hemoglobinemia can be detected by simply inspecting the supernatant of a carefully drawn and centrifuged sample of venous blood. With severe hemolysis, hemoglobinuria is also observed. When both hemoglobinuria and hemoglobinemia are present, intravascular hemolysis is certain. Over the next 24 hours, the serum haptoglobin level falls to undetectable levels and, if no further hemolysis occurs, only gradually recovers over the next 2 to 3 days. It is therefore possible to use the haptoglobin level, as well as measurements of the plasma methemalbumin and urinary hemosiderin, to understand the duration of the intravascular hemolytic process. Although the haptoglobin level recovers quite quickly after hemolysis is stopped, elevated levels of plasma methemalbumin and detectable amounts of hemosiderin in the urine are present for 7 to 10 days.

The presence of hemoglobinuria alone usually does not indicate intravascular hemolysis. It may be the result of bleeding from the kidney or lower urinary tract with lysis of red cells in the bladder caused

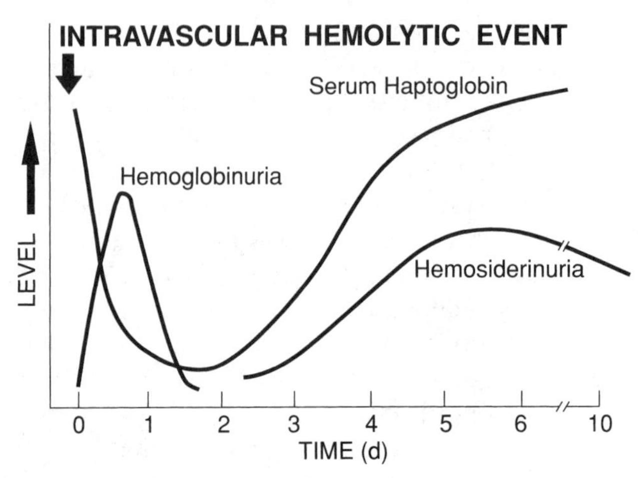

Figure 49. Indicators of acute intravascular hemolysis. Within a few hours of an acute hemolytic event, free hemoglobin is cleared from plasma, and the serum haptoglobin falls to undetectable levels. Hemoglobinuria ceases soon after this. If no further hemolysis occurs, the serum haptoglobin level recovers, and methemalbumin disappears within several days. Measurement of the urinary hemosiderin can provide more lasting evidence of the hemolytic event.

by the hypotonicity of the urine. If bleeding from the kidney or lower urinary tract is suspected, a urine specimen should be centrifuged and the sediment examined for intact red cells (hematuria). Myoglobinuria can also mimic hemoglobinuria. However, the plasma will not be pink because the small myoglobin molecule is rapidly cleared by plasma into the urine. Urine dipstick methods available for detecting hematuria do not distinguish between myoglobin and hemoglobin. Furthermore, the dipstick methods are extremely sensitive to even small numbers of intact red cells, making the tests overly sensitive but of low specificity. Myoglobin and hemoglobin can be differentiated spectroscopically. Finally, other heme pigments that can color the urine include porphobilinogen, dipyrrols, and urobilinogen. These pigments are usually colorless at the time when the urine is voided but darken if the urine is allowed to stand.

The most sensitive indicator of chronic intravascular hemolysis is hemosiderinuria. In fact, with milder degrees of intravascular hemolysis, hemosiderinuria can be present without hemoglobinuria, reflecting the capacity of renal tubular cells to reabsorb hemoglobin and convert it to hemosiderin. Chronic conditions such as paroxysmal nocturnal hemoglobinuria and red cell fragmentation caused by vasculitis or a prosthetic heart value can result in sufficient hemosiderinuria to produce iron deficiency.

Acute extravascular hemolysis, in which there is little or no release of hemoglobin into plasma or urine, can be more difficult to recognize. A sudden drop in hemoglobin and an increase in the indirect bilirubin and LDH levels should alert the physician to the possibility of extravascular hemolysis. If the hemolysis is ongoing for more than 1 week, a brisk reticulocyte response should also be present. This will not be true if the patient is evaluated within the first few days after a limited hemolytic event. Severe renal dysfunction or an inflammatory process can also impair the marrow's reticulocyte response. The presence of liver disease can make the diagnosis difficult by affecting both the bilirubin and LDH levels.

Self-limited extravascular hemolytic events are most frequently seen in patients with a defect in aerobic glycolysis such as G6PD deficiency or with an unstable hemoglobin.[72] Under these circumstances, the red blood cell cannot adequately defend against oxidative damage to its hemoglobin by a drug. An episode of extravascular hemolysis may also occur as a result of an autoimmune process involving either a cold or warm antibody. In patients receiving multiple transfusions, an anamnestic immune response to a minor blood group antigen (most often one from the Rh, Kell, Duffy, or Kidd systems) can result in the sudden destruction of cells within a few days of a transfusion. These patients experience little in the way of symptoms or signs that would lead to the appreciation of the hemolytic event. Rather, the sudden loss of red cells is appreciated from a rapid fall in the hemoglobin level and, in the case of an autoimmune or blood transfusion reaction, from the appearance of a positive direct or indirect antiglobulin test. Immediately after a hemolytic episode, G6PD-deficient patients can be most difficult to diagnose. The red blood cells remaining in circulation have relatively normal levels of G6PD, so that a simple screening test for the enzyme defect is normal. In this situation, a repeat test after the patient recovers or a G6PD assay following incubation of the patient's blood will be necessary to make the diagnosis.

Chronic Hemolysis

The Clinical Picture

Chronic extravascular hemolysis generally produces a well-compensated chronic hemolytic anemia. The typical erythropoietic profile is summarized in Table 25. Most patients demonstrate a mild to moderately severe anemia with a reticulocyte production index of greater than three times the basal level. Patients with se-

Table 25 • **Hemolytic Anemia: Erythropoietic Profile**

Smear/Red Cell Indices
- Normocytic or macrocytic, often with a unique morphologic abnormality (spherocytes, elliptocytes, fragmentation, "bite cells")
- MCV = 80–140 fL, shift macrocytosis, proportionate to the degree of anemia and iron supply

Reticulocyte Index
- Greater than 3.0 (3–6 times basal level)
- Polychromasia prominent

Marrow E:G Ratio
- Greater than 1:1
- Normoblastic erythroid hyperplasia

Indirect Bilirubin/LDH Levels
- Increased

vere congenital hemolytic anemias can develop a greatly expanded erythroid marrow and a reticulocyte index that reaches five to six times basal. The ability of the marrow to compensate for chronic hemolysis is also striking. Patients with hereditary spherocytosis, a congenital defect in the structure of the red cell membrane, can fully compensate for the increased rate of red cell destruction, so that their hemoglobin is within the normal range.

Red cell destruction by the reticuloendothelial system (extravascular hemolysis) predominates in patients with lifelong chronic hemolytic anemias. Patients who have chronic intravascular red cell fragmentation (artificial heart valves) or paroxysmal nocturnal hemoglobinuria can, however, demonstrate a component of intravascular hemolysis as well. The importance of this is the potential for urinary iron loss caused by the hemoglobinuria, hemosiderinuria, or both. Over time, even relatively small amounts of iron loss into urine can produce an iron-deficiency state, which then suppresses red cell production. The increased erythropoiesis of a chronic hemolytic anemia may also be suppressed by inflammation. Cytokines released as a part of the inflammatory response block the release of iron from the reticuloendothelial cell and inhibit the erythropoietin response. With an acute infection, the rate of red cell destruction may also be accelerated. The combination of more severe hemolysis and a temporary suppression of erythropoiesis can dramatically worsen the anemia. This worsening is commonly seen in patients with sickle cell anemia and hereditary spherocytosis and is referred to as a *hemolytic crisis.*

Patients with chronic hemolytic anemias have an increased folate requirement because of the high level of erythroid marrow proliferation. Therefore, any illness associated with a reduction in the dietary intake of folate can result in the rapid appearance of ineffective erythropoiesis and a fall in the reticulocyte production index. If the folate deficiency continues over several weeks, the patient develops a macrocytic anemia with a megaloblastic bone marrow. Chronic hemolytic anemia patients are also at risk for episodes of temporary erythroid marrow aplasia. This is most often associated with a parvovirus infection and involves an actual shutdown of erythroid precursor proliferation despite adequate levels of erythropoietin stimulation. It is reversible, although the patient requires transfusion support during an aplastic crisis that lasts for more than a few days.

The role of the spleen is also important in determining the clinical picture of a chronic hemolytic anemia. Many patients demonstrate a modest increase in splenic size, reflecting an expansion of the splenic red cell pulp. If splenic enlargement becomes pronounced, it can result in excessive trapping of platelets and granulocytes, leading to pancytopenia in some patients. Marked splenomegaly from any cause can result in severe pancytopenia and a further acceleration of red cell destruction (hypersplenism). In this situation, the spleen not only removes the patient's abnormal red cells from circulation at an accelerated rate, but also destroys transfused, normal red blood cells.

Chronic hemolytic anemias can result from an environmental factor or a defect in any of the major components of the red blood cell—its membrane, metabolic machinery, or hemoglobin structure (Table 26). From the standpoint of clinical frequency, the most common environmental and membrane defects include mechanical fragmentation, autoimmune damage, and the congenital membrane defect of hereditary spherocytosis. In black and Asian populations, defects in hemoglobin structure (sickle cell anemia, Hb C and Hb SC disease, and combinations of Hb S with thalassemia) are the most common. In Mediterranean and African populations, phosphogluconate pathway defects (G6PD and glutathione–GSH–deficiency) have a high clinical incidence.[72] At the same time, there are a large number of other metabolic and hemoglobin structural defects that produce rarer congenital chronic hemolytic anemias.

Table 26 • **Causes of Extravascular Hemolysis**

Environmental Disorders
- Infections (malaria, bartonellosis, mycoplasma, infectious mononucleosis)
- Drug-induced (oxidant drugs in G6PD-deficient patients, immune drug reactions)
- Autoimmune (transfusion reactions, warm and cold autoimmune hemolytic anemias)
- Hemolytic-uremic syndrome (HUS)
- HELLP (hemolysis, elevated liver enzymes, low platelets)

Membrane Defects
- Hereditary spherocytosis
- Hereditary elliptocytosis
- Paroxysmal nocturnal hemoglobinuria
- Acanthocytosis

Metabolic defects
- Phosphogluconate pathway (G6PD, GSH deficiencies)
- Embden-Meyerhof pathway (pyruvate kinase, hexokinase, other enzyme deficiencies)

Hemoglobin structural defects
- Hemoglobinopathies (Hb S, Hb C, and α thalassemias)
- Unstable hemoglobins

Differential Diagnosis

The differential diagnosis of a chronic hemolytic anemia can be difficult because there are so many possible causes. There is no easy way around this problem. Some of the hemolytic states do present with a unique clue, such as the painful vasocclusive crises of sickle cell anemia or the hemoglobinuria associated with cold exposure in a patient with a high-titer cold agglutinin. More often, the physician must search among a number of general clues for a diagnostic combination. Several areas of inquiry can be helpful:

- **Is the anemia associated with an acute illness or a complicating disease?** If there seems to be a strong relationship between another illness and the onset of the anemia, a bacterial or parasitic infection, an acquired autoimmune process, or a defect in the metabolic machinery of the red blood cell is likely.
- **Does a drug seem to be responsible?** Here again, an autoimmune process or a defect in metabolic machinery is more likely.

- **Does the anemia appear to be hereditary?** An anemia that can be traced through several members of the family and/or that has been present throughout the patient's life is almost certainly a congenital hemolytic anemia.
- **What is the patient's race?** Hemoglobinopathies and enzyme deficiencies resulting in chronic hemolytic anemias are genetic disorders that follow clear racial and geographic lines.

Based on these questions, the probability of a specific hemolytic anemia may be very high, and a single laboratory test targeted at a specific defect can be diagnostic (Table 27). For example, detection and diagnosis of a hemoglobinopathy are possible with a high degree of certainty by hemoglobin electrophoresis. Similarly, the diagnosis of hereditary spherocytosis is usually possible using the osmotic fragility and incubated autohemolysis tests. G6PD deficiency may be detected with a simple G6PD screening test. However, the definitive detection of most of the rarer defects in metabolic machinery requires the expertise of a hemolytic anemia reference laboratory.

When the clinical picture does not suggest a specific diagnosis, a battery of laboratory tests may be needed to identify the cause for the

Table 27 • **Laboratory Tests for Hemolysis**

Intravascular
Smear inspection for fragmentation
Plasma inspection for free hemoglobin
Plasma haptoglobin level
Urine inspection for hemoglobin
Urine hemosiderin
Extravascular
Coombs (antiglobulin) test/cold agglutinin titer
Hemoglobin electrophoresis
Heinz body preparation
G6PD screen
Isopropanol denaturation test
Osmotic fragility test
Incubation hemolysis test
Sucrose hemolysis test
Enzyme assays and P_{50}

hemolysis. There is no absolute, right way to sequence the workup. One diagnostic algorithm for the workup of a chronic hemolytic anemia is shown in Figure 50. The need to rule out intravascular hemolysis as a first step in diagnosis has already been discussed but should be emphasized. Tests for the detection of intravascular hemolysis include a careful inspection of the blood smear for red blood cell fragmentation and inspection of plasma and urine specimens for free hemoglobin and urine hemosiderin. Severe, life-threatening intravascular hemolysis is associated with hemoglobinemia and hemoglobinuria. With less severe, chronic intravascular hemolysis, the inspection of the smear for red blood cell fragmentation and the urine for hemosiderinuria is more helpful.

In chronic extravascular hemolytic anemia, the CBC and red blood cell morphology may point the way to the diagnosis. Several distinctive red blood cell shapes serve as diagnostic markers for common chronic hemolytic anemias (see Fig. 30). For example, red blood cell fragmentation may result from thermal burns, wide-

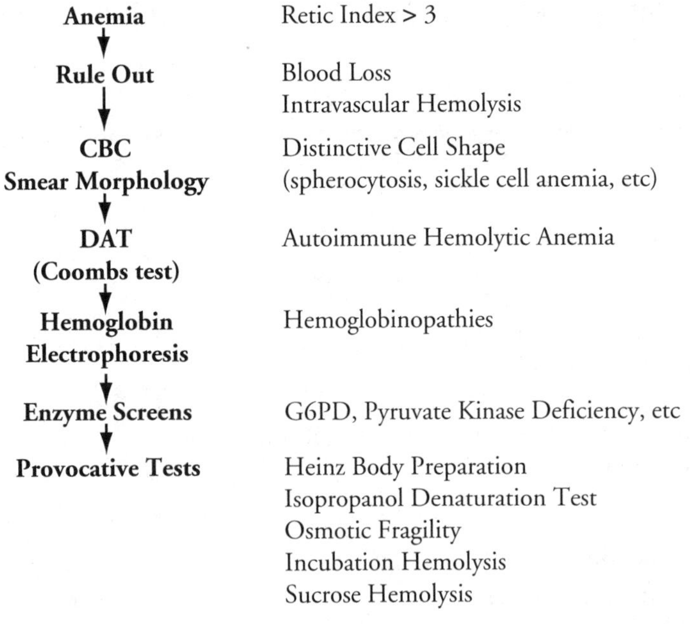

Figure 50. Stepwise evaluation of a chronic anemia with a high reticulocyte index. After chronic blood loss and intravascular hemolysis are ruled out, laboratory studies can be organized according to the red blood cell morphology and the relative frequency of extravascular hemolytic disorders in the target population.

spread malignancy, and mechanical damage from an artificial heart valve. Malaria can be diagnosed by direct visualization of the parasite in red blood cells. Unique red blood cell shapes, such as elliptocytes and sickle cells, by themselves are diagnostic. Spherocytes are seen in patients with hereditary spherocytosis and as a result of red cell membrane damage from IgG autoantibodies. Red cell targeting, involving more than 50% of circulating cells, is seen in patients with Hb C disease and patients with obstructive jaundice.

Interpretation of the CBC and smear morphology can, however, be challenging. One misleading morphologic change in patients with hemolytic anemias is the appearance of macrocytosis secondary to the reticulocytosis. An increased level of erythropoietin stimulation results in the release from the erythroid marrow of young "shift" reticulocytes. These cells have a volume of 140 to 160 fL. As their numbers increase in circulation to uncorrected reticulocyte counts of 10% or more, prominent polychromasia on the peripheral blood smear and an increase in the MCV and RDW are seen. In patients with a severe autoimmune hemolytic anemia or a congenital hemolytic anemia, uncorrected reticulocyte counts of greater than 20% to 30% are not uncommon. This will increase the MCV to levels of 110 to 130 fL, a range usually associated with the macrocytic, megaloblastic anemias.

Because the increase in the MCV should match the level of reticulocytosis, a normal MCV in some patients requires an explanation. Patients with hereditary membrane abnormalities, especially hereditary spherocytosis, and certain hemoglobinopathies (Hb S and Hb C disease) generally have a normal MCV (80 to 100 fL), despite the presence of a high reticulocyte count. The absence of macrocytosis in these patients reflects a mild hemoglobin synthesis defect and/or a progressive loss of cellular material as cells circulate. The latter is most prominent in patients with hereditary spherocytosis, in which circulating red blood cells lose cellular water, hemoglobin, and portions of the red cell membrane. This loss produces a cell that is both microcytic and overly dense, as reflected by an increase in the MCHC to levels in excess of 36%. The combination of many small, dense spherocytes (microspherocytes) on the peripheral blood smear and a high MCHC is unique enough to be diagnostic for hereditary spherocytosis.

When patients with congenital hemolytic anemias demonstrate an MCV below 80 fL, they must be evaluated for an associated iron deficiency or the presence of a genetic disorder with impairment in hemoglobin synthesis, such as one of the thalassemias that presents as a hemolytic anemia (Hb H disease). Finally, an increase in the MCV out of proportion to the patient's reticulocytosis must always raise the possibility of folic acid deficiency. In patients with very high red cell production levels, folate supplementation is generally required to prevent the development of macrocytosis. This increases the likelihood of a complicating macrocytosis in any patient with a long-standing hemolytic anemia.

Environmental Disorders

Red blood cells can fall victim to a number of systemic disease states, resulting in an acute or chronic hemolytic anemia. The mechanism of destruction can involve either the membrane, the cell's metabolic machinery, or the structural integrity of the hemoglobin. Depending on which is involved, red blood cell morphology may provide a clue to the diagnosis. However, it is the overall clinical picture that usually alerts the clinician to the cause of the hemolysis.

A number of bacterial, viral, and parasitic diseases produce hemolytic anemias. The most prominent of these include clostridial sepsis, the hemolytic uremic syndrome of *Escherichia coli* gastroenteritis, and the hemolysis caused by direct red cell invasion by parasites such as malaria and *Bartonella.* Mycoplasma pneumonia and infectious mononucleosis are also associated with hemolytic events secondary to the production of immunoglobulin M (IgM) antibodies.

The drugs used to treat infections can also be the culprit. Oxidant drugs, including many of the antibiotics, the antimalarials, and some chemotherapeutic drugs cause an acute, self-limited hemolysis in patients with the most common form of G6PD deficiency, the X-linked A-minus variant.[72] This is most commonly seen in African American men and boys. A patient with the Mediterranean variant of G6PD deficiency or GSH deficiency can experience a more severe, life-threatening hemolysis. However, this is a much less common event.

Drugs can cause an immune hemolytic anemia. Mechanisms include the induction of antibodies to the red blood cell membrane structure, nonselective binding of the drug to the membrane with subsequent destruction by a drug-directed antibody, or the formation of a drug-antibody complex with hemolytic potential. The penicillins, quinidine, and alpha methyldopa are the most common drugs associated with red cell hemolysis. In the case of the penicillins, the destruction of the red blood cell is a side effect of an antibody specific to the drug. Quinidine represents the best example of a drug (haptene)-antibody interaction, while alpha methyldopa destroys red cells by inducing an autoantibody to the Rh antigens of the membrane.

Immune destruction of red cells can result from an autoimmune disorder, collagen-vascular disease, or lymphopoietic malignancy. The ready availability of laboratory tests to identify the presence of IgG, IgM, and complement on the red blood cell membrane helps with diagnosis of this group of conditions, called the *autoimmune hemolytic anemias* (AIHAs). The warm-autoantibody type of AIHA involves an IgG antibody that may or may not bind complement and may or may not show specificity for a blood group antigen such as the Rh locus. Typically, IgG autoimmune red blood cell destruction is purely extravascular. It is most frequently seen in patients with collagen-vascular disease, especially lupus erythematosus, and lymphoproliferative disorders such as chronic lymphocytic leukemia. As previously mentioned, the drug alpha methyldopa is capable of inducing a mild autoimmne hemolytic anemia by inducing an IgG antibody to the Rh locus of the red cell. Many patients with this type of anemia have no obvious cause for the disorder, which is therefore labelled idiopathic.

The other clinical form of AIHA involves a cold-reacting IgM antibody that tends to produced red cell agglutination and both intravascular and extravascular hemolysis.[73,74] Many of the IgM cold antibodies have I or i red cell antigen specificity. They may be temporarily evoked by infections (mycoplasma pneumonia and infectious mononucleosis) or be a chronic problem in patients with lymphoproliferative disorders. Similar to the idiopathic form of warm-antibody disease, cold-agglutinin disease can occur without evidence of another disease process. The severity of the hemolytic process tends to be limited by the temperature specificity of the antibody and its complement-binding characteristics. Some patients can show dramatic intravascular hemolysis when exposed to the cold. Others show a relatively stable level of chronic extravascular hemolysis. In this case, the patient's red cells appear relatively resistant to destruction because complement on the red cell membrane is converted to C3d, its inactive form. In contrast, freshly transfused red blood cells may be rapidly destroyed, especially if fresh complement is administered as a part of the red cell transfusion.

Some patients with an AIHA do not have a positive direct antibody test (DAT, or Coombs, test) because the amount of antibody attached to the red cell surface is too little to be detected. This is a reflection of the poor sensitivity of the routine DAT test, which cannot detect fewer than 500 antibody molecules per cell. More sensitive research techniques are available that can be used to identify individuals with smaller antibody loads, but these require the expertise of a special hematology laboratory. From a practical standpoint, a therapeutic trial with steroids can be used to aid in the differential diagnosis. If a patient responds to treatment with steroids, an autoimmune disorder not recognized by the standard DAT test may be inferred.

Rapid red blood cell destruction also accompanies systemic diseases that damage the vasculature. Hemolytic-uremic syndrome (HUS) and thrombotic thrombocytopenia purpura (TTP) are examples of systemic conditions in which a defect in the endothelial cell lining appears to affect red cell survival. In patients with hemolytic-uremic syndrome, the severity of the hemolysis can be impressive, even dominating the clinical presentation. In patients with TTP, the hemolysis component is usually overshadowed by the severe thrombocytopenia and evidence of neurologic damage. In both conditions, plasmapheresis and infusions of fresh plasma depleted of von Willebrand factor appear to have a salutary affect on both the hemolysis and rapid destruction of platelets. Other conditions associated with red cell hemolysis include widespread malignancy and eclampsia or preeclampsia. Here again, distortions of the vasculature may well be responsible for the red cell destruction.

Membrane Defects

Inherited defects in the protein structure of the red blood cell membrane (hereditary spherocytosis and hereditary elliptocytosis) produce lifelong, well-compensated hemolytic anemias.[75] Acanthocytosis with hemolysis can occur as a complication of severe cirrhosis or pancreatitis or as a prominent component of inherited abetalipoproteinemia. Similarly, stomatocytosis occurs in patients with cirrhosis and neoplasms as well as an inherited defect in the expression of Rh antigen on the red cell membrane. Paroxysmal nocturnal hemoglobinuria is an example of an acquired membrane defect secondary to a clonal malignancy of the red cell line.

Hereditary spherocytosis is the most common of the membrane disorders. It is inherited as a dominant trait and presents as a well-compensated hemolytic anemia with spherocytosis on the peripheral blood smear, an increase in the MCHC, and mild to moderate splenomegaly. The severity of the expression of the defect over several generations of an involved family can vary greatly. Some patients have quite severe disease and complain of a functional impairment; others have few or none of the findings associated with hereditary spherocytosis and appear clinically well. In the latter, disease can be very difficult to diagnose.

The underlying defect in hereditary spherocytosis is an inherited deficiency in spectrin and/or ankyrin, the principal proteins of the submembrane structure of the red cell.[75] Although this structural deficiency can be detected by protein analysis, the clinical diagnosis of hereditary spherocytosis is usually made with two provocative tests, the incubated osmotic fragility test and the autohemolysis test (Figs. 51 and 52). The osmotic fragility test subjects the red blood cell to the stress of increasingly hypotonic salt solutions. Circulating spherocytes demonstrate an increased susceptibility to hemolysis, an abnormality that can be magnified by first

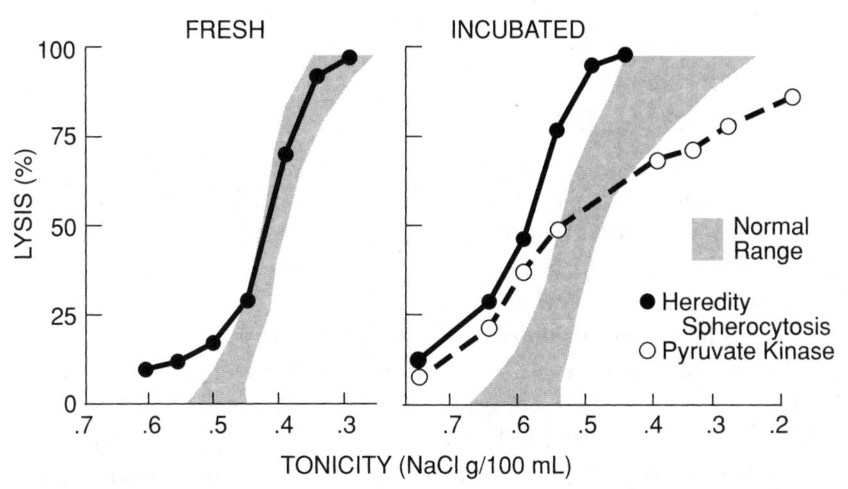

Figure 51. Osmotic fragility test. This test may be used to identify patients with defects in red cell membrane (i.e., hereditary spherocytosis) or intracellular metabolism (i.e., pyruvate kinase deficiency). The osmotic fragility test involves subjecting the red cells to stress of increasingly hypotonic salt solutions. When the test is performed on a fresh blood sample, there is displacement of the hereditary spherocytic curve to the left, representing increased susceptibility to hemolysis. Often this is limited to a small tail of cells that are unusually susceptible to lysis. After incubation, the defect is magnified. In contrast, pyruvate kinase–deficient blood shows a mix of sensitive cells and cells resistant to osmotic lysis.

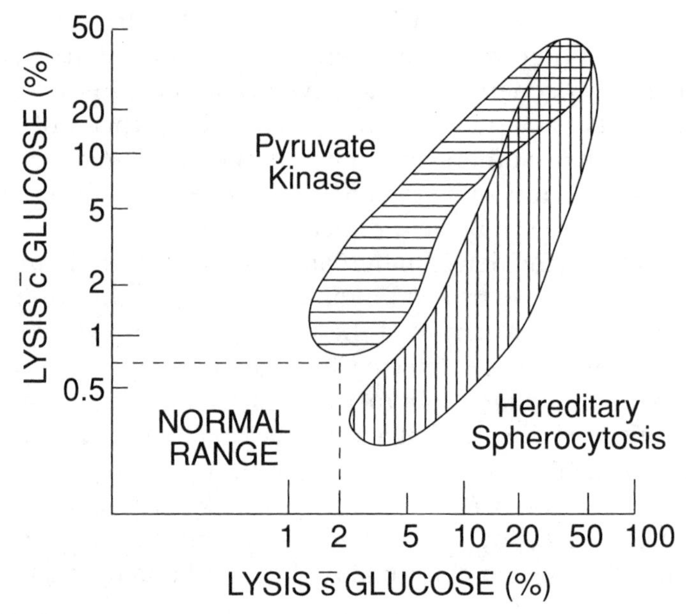

Figure 52. The autohemolysis test provides a further measure of cell resistance to hemolysis. Pyruvate kinase–deficient blood demonstrates an abnormal rate of hemolysis that is independent of the presence or absence of glucose in the incubation media. In contrast, the blood from a patient with hereditary spherocytosis shows more marked hemolysis when glucose is absent.

incubating the blood sample. The autohemolysis test, with and without added glucose, can also help detect milder disease. Typically, the red blood cells of the patient with hereditary spherocytosis show a marked increase in cell lysis, which is magnified by incubation without added glucose.

Hereditary elliptocytosis is easily diagnosed from the morphology of red blood cells on the peripheral smear; more than one half of the circulating red cells are of a uniform elliptical or oval shape. It is an inherited defect in the composition of spectrin molecules in the membrane. Some, but not all, patients with hereditary elliptocytosis demonstrate a mild hemolytic anemia. Acanthocytosis is also diagnosed from the peripheral blood smear. Both the number of cells involved and the severity of the hemolytic anemia depend on the cause of the abnormality. A few acanthocytes in patients with cirrhosis or pancreatitis are usually not associated with a clinically important hemolytic process.

Paroxysmal nocturnal hemoglobinuria (PNH) patients can

present with a hemolytic anemia or a severe pancytopenia, reflecting the fact that PNH is a clonal malignancy that evolves into a dysplastic or acute leukemic state.[76] The red blood cell hemolysis reflects the acquired defect in the expression of complement regulatory proteins on the red cell membrane. This makes the involved cells extremely sensitive to complement-mediated lysis; the classic PNH patient presents with a history of recurrent episodes of nocturnal hemoglobinuria. With time, sufficient iron can be lost in the urine to suppress red blood cell production and dampen these episodes, whereas iron therapy can worsen the hemoglobinuria.

Provocative screening tests that are useful in diagnosing PNH include the sugar-water test and the HAM acid hemolysis test. Both of these involve stimulating complement activation by the indirect pathway to stress the red blood cells. More recently, it has become possible to assay directly for the involved membrane proteins (CD55 and CD59) using monoclonal antibodies and flow cytometry. In patients with a markedly dysplastic marrow, the provocative tests can be less reliable. Patients may need to be repeatedly tested over a period of time to detect the abnormality.

Metabolic Defects

The metabolic pathways of the red blood cell, especially the phosphogluconate and Embden-Meyerhof pathways, are fertile ground for inherited defects of individual enzymes, which increase the hemolytic potential of the red blood cell. The common A-minus variant of G6PD deficiency, seen in African American populations, is a model for the red cell instability caused by one of these defects.[72] These patients do not have a chronic hemolytic anemia, but they are at risk for a hemolytic event when exposed to an oxidant chemical or drug. Other genetic variants of the phosphogluconate pathway, the Mediterranean form of G6PD deficiency and severe GSH deficiency, are associated with more severe and even chronic hemolytic anemias. As for defects in the Embden-Meyerhof pathway, pyruvate kinase deficiency and hexokinase deficiency are perhaps the most common; they are associated with a moderately severe hemolytic anemia.

The osmotic fragility and autohemolysis tests (see Figs. 51 and 52) also can be used to screen for defects in the Embden-Meyerhof pathway. Pyruvate kinase deficiency shows osmotic fragility and autohemolysis patterns that are quite different from those seen wither hereditary spherocytosis. In addition, patients with defects in metabolic machinery do not show spherocytes or elliptocytes. As another clue, pyruvate kinase–deficiency patients demonstrate a persistence of RNA and reticulum in circulating reticulocytes, resulting in uncorrected reticulocyte counts in excess of 50% to 70%. The diagnosis of the specific enzyme defect may be possible with a simple enzyme screening test (pyruvate kinase deficiency) or require the expertise of a hematology reference laboratory. In evaluating such a patient, it may be worthwhile to study the oxygen dissociation curve or measure the red cell 2, 3-DPG level. In enzyme defects located in the glycolytic sequence before the production of 2, 3-DPG, red cells have decreased 2, 3-DPG and a left shift of the oxygen dissociation curve (hexokinase, glucose-6-phosphate isomerase, and phosphofructokinase deficiencies).When the lesion is below this level (including pyruvate kinase deficiency), the 2,3-DPG level is high, and the oxygen dissociation curve is shifted to the right.

Hemoglobin Structural Defects

Patients with hemoglobinopathies often present with a chronic hemolytic anemia. The severity of the anemia can vary considerably, depending on the nature of the genetic defect and whether the patient is a heterozygote, homozygote, or compound heterozygote. The common hemoglobinopathies in the United States are listed in Table 28. Most involve the mutation of a single amino acid in one of the globin chains. In sickle cell disease, the abnormality is a substitution of valine for glutamic acid in position 6 of the β-globin chain. Hb C results from a substitution of lysine for glutamic acid in the same locus, whereas Hb D involves a substitution of glutamine for glutamic acid in position 121 of the β chain.

Sickle cell anemia is the most common and best studied of the hemolytic anemias resulting from a hemoglobin structural defect. Sickle hemoglobin, Hb S, is unstable in its deoxy hemoglobin form. In homozygous patients, crystal-like tactoids of Hb S distort the architecture of the red blood cell, producing the distinctive crescent-moon-shaped sickle cell (see Fig. 30). The rigidity of sickled cells interferes with their ability to pass through small arterioles and capillaries. Log jams of cells plug small vessels and lead to localized tissue infarction or, when extensive, to disseminated intravascular coagulation.

The physical behavior of sickled hemoglobin defines the clinical presentation. Patients present early in childhood with symp-

Table 28 • **Common Hemoglobinopathies**

Hemoglobin S Disorders
Sickle cell anemia (SS)
Sickle cell anemia with high F levels
Sickle cell trait (SA)
SC disease
S/β thalassemia
S/Lepore hemoglobin
SD disease
S/O-Arab disease
Hemoglobin C Disorders
Homozygous C disease
Hemoglobin C trait
Unstable Hemoglobins
Hemoglobin Zurich
Hemoglobin Koln

toms and signs of vaso-occlusive disease of key organs including the marrow, spleen, kidney, and central nervous system. Vaso-occlusive episodes are recognized clinically as recurrent painful crises with localized or diffuse bone and joint pain. Splenic infarction occurs early in childhood, with a loss of splenic function. Similarly, infarction of the renal medulla results in a loss of concentrating ability and, in some patients, episodes of gross hematuria. Infarctions of the lung and central nervous system can be life-threatening. This occurs most frequently during the second or third decade of life.

The clinical severity of Hb S disease is directly related to the inheritance pattern of the various normal and abnormal hemoglobins. When only a single gene is involved (the heterozygous form of Hb S disease, called *sickle cell trait*), patients do not exhibit painful crises or organ damage and have normal life spans. Homozygous Hb S patients (sickle cell anemia) can also demonstrate a variable clinical picture depending on the level of intracellular Hb F. Hereditary persistence of Hb F production, as seen in patients from the Middle East, results in mild or minimal disease. Iron deficiency or the combined inheritance of thalassemia can also reduce the severity of the disease, perhaps by decreasing the hemoglobin concentration (MCHC) of the red blood cell. Combined heterozygotes with Hb C and β° thalassemia (Mediterranean populations) tend to exhibit more severe disease, whereas those with the combination of Hb S with β^+ thalassemia (African populations) demonstrate only a mild to moderate hemolytic anemia and splenomegaly. The latter much more resemble patients with sickle cell trait.

The laboratory diagnosis of Hb S, Hb C, and combinations with β thalassemia is possible using the techniques of hemoglobin electrophoresis and measurements of Hb A_2 and Hb F. The electrophoretic patterns for the most common hemoglobinopathies are shown in Figure 53. Hb S and Hb C tend to migrate more slowly than Hb A or Hb F. Therefore, the patterns for sickle cell trait, sickle cell disease, S/C disease, and C trait are easily distinguished. The ratio of Hb S to Hb A is also important. Patients with sickle cell trait have an Hb S:Hb A ratio of about 40:60 because the production of Hb S is slightly less efficient than that of Hb A. In contrast, com-

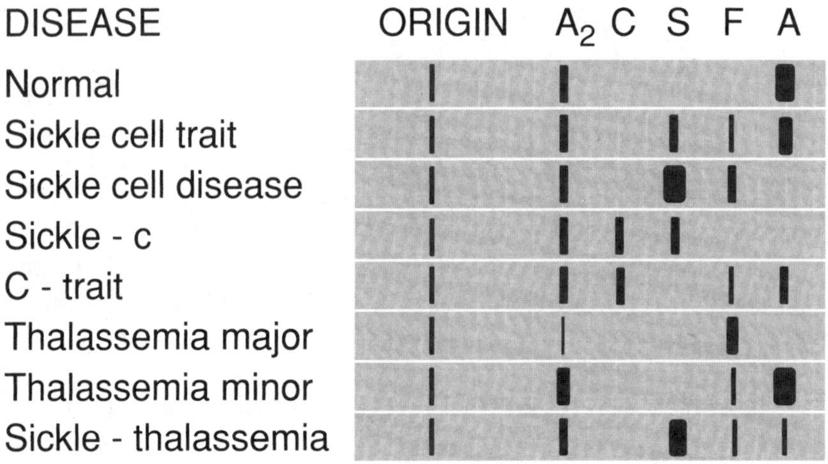

DISEASE — ORIGIN A_2 C S F A
Normal
Sickle cell trait
Sickle cell disease
Sickle - c
C - trait
Thalassemia major
Thalassemia minor
Sickle - thalassemia

Figure 53. Cellulose acetate electrophoretic patterns for common hemoglobinopathies. In the normal individual, 97% of the hemoglobin found in circulating red cells is Hb A. Only small amounts of Hb F and A_2 are detectable. Patients with sickle cell trait and sickle cell disease show increased amounts of Hb S with a corresponding decrease in Hb A. Patients with sickle cell disease may show a variable increase in Hb F. Sickle cell-c patients show an increase in the hemoglobin band in the A_2 c position, which in fact is Hb C. Patients with sickle cell thalassemia show increases in the A_2 and S bands and a decrease in the Hb A band, which is more marked than that observed in sickle cell trait patients. Patients with thalassemia major show a decrease in Hb A and a marked increase in Hb F. In contrast, the patient with β thalassemia minor shows only a slight increase in Hb F together with an increase in Hb A_2.

pound heterozygote patients with Hb S and β^+ thalassemia (African type) have a reverse hemoglobin Hb S: Hb A ratio of 60:40. This reflects the impact of the β^+ thalassemia mutation on Hb A production. Patients with Hb S β° thalassemia (Mediterranean type) have an electrophoretic pattern that mimics the pattern of sickle cell disease, with an Hb S:Hb A ratio of 100:0. Distinguishing features include a slight increase in Hb A_2 and a more severe microcytosis in the Hb S β° thalassemia patients. In assessing the Hb F level in these patients,the acid elution slide test for intracellular Hb F can help differentiate patients with thalassemia from those with a hereditary persistence of Hb F production. The latter tend to show a more uniform presence of Hb F in circulating red blood cells.

In addition to these common clinical hemoglobinopathies, more than 400 hemoglobin structural variants have now been described. Some of these result in a physically unstable molecule that precipitates within the cell, producing a hemolytic anemia. Others are associated with a change in the hemoglobin-oxygen affinity, causing either polycythemia or anemia. These conditions are not associated with a characteristic morphologic change. Rather, they are diagnosed using screening tests such as the isopropanol denaturation and heat stability tests and brilliant cresyl blue and methyl violet supervital stains to identify denatured hemoglobin precipitates within the cells (Heinz bodies). Definitive diagnosis of the hemoglobin variant then requires the use of electrophoretic techniques and globin chain amino acid sequencing.

Chapter

4 Polycythemia

The terms *polycythemia* and *erythrocytosis* are both used clinically to describe an abnormally high hemoglobin or hematocrit measurement. They are not truly interchangeable terms. Erythrocytosis is the more accurate descriptor for an increased red blood cell mass without changes in other cell lines. However, polycythemia is the more popular term and is used by most clinicians to describe any condition that results in a red cell mass increase. Even the rise in hematocrit that is solely the result of a decrease in plasma volume is referred to as *relative polycythemia.*

The likelihood that a patient has an increased red blood cell mass is a function of just how far the hemoglobin or hematocrit exceeds the expected normal range. The upper limit of a "normal" hemoglobin concentration varies according to age, sex, and the altitude at which the patient lives, but an arbitrary upper limit (at sea level) can be set at 17.5 g/dL for men and 16 g/dL for women. Just as with anemia, there is an overlap between the physiologic norm and polycythemia. The probability of any hemoglobin value being normal or abnormal depends on the incidence of polycythemia in the population. For example, although the incidence of polycythemia from any cause is very low for the general population, patients with chronic obstructive pulmonary disease can be expected to have a higher incidence of hypoxic polycythemia.

The same factors that interfere with the accurate diagnosis of an anemia tend to confound the measurement of high hemoglobin values. The state of fluid balance holds special importance. Since both hematocrit and hemoglobin levels reflect the amount of diluent plasma, moderate elevations of hemoglobin to 18 to 20 g/dL in men can result simply from a plasma volume reduction. However, a persistent hemoglobin value of 20 g/dL or higher in men or 18 g/dL or higher in women is almost always associated with a true increase in red cell mass. In this situation, direct measurements of blood volume are unnecessary.

Any illness that suppresses the red blood cell production level of a patient with polycythemia lowers or normalizes the hemoglobin. Iron deficiency secondary to chronic blood loss is a common complication in these patients and lowers the hemoglobin level. The very early appearance of microcytosis and hypochromia, before the hemoglobin falls below the normal range, is an indication that the patient has an underlying polycythemia complicated by iron deficiency.

Clinical Signs and Symptoms

Polycythemia, like hypertension, can produce remarkably few symptoms. At hemoglobin levels in excess of 18 g/dL (hematocrit greater than 54%), whole blood viscosity increases, especially with the low flow rates encountered in the microvasculature. This increased viscosity represents a threat to tissue oxygenation.[77] Cerebral circulation and oxygen supply are significantly reduced, and lactate levels in exercising individuals increase. This is true even for patients with relative polycythemia, in whom the increase in the hematocrit is caused solely by a reduction in the plasma volume. Patients complain of headache, general malaise, and easy fatigability. They may also be mildly hypertensive. It is not always clear whether the symptoms are related to increased viscosity or to an underling disease producing the polycythemia, but phlebotomy frequently provides symptomatic relief.[78]

Extremely high hemoglobin levels (greater than 20 g/dL at sea level) are associated with both an increase in blood viscosity and total blood volume. The resulting increased cardiac output and cardiac work may precipitate congestive heart failure in an individual with borderline cardiac function. In addition, the patient is at increased risk for both bleeding and thrombotic complications. A consumptive coagulopathy may be seen in hypoxic patients with marked polycythemia secondary to congenital heart disease.

Physical findings related to polycythemia are variable and may not alert the clinician to the diagnosis. When the hemoglobin level exceeds 20 g/dL, the patient appears plethoric, with bloodshot eyes and conjunctiva. Vessels of the ocular fundus and mucous membranes may appear dilated and somewhat cyanotic. If the polycythemia is secondary to marked arterial desaturation, cyanosis will be prominent. With lower hemoglobin levels, plethora may or may not be prominent. A patient's skin color depends a great deal on the thickness of the skin, its pigmentation, and the amount of cutaneous blood flow.

Approach to Diagnosis

The patient's history can provide important clues to the etiology of the polycythemia. Special attention should be given to the altitude at which the patient lives and other environmental factors that can affect arterial oxygen saturation, such as pulmonary dysfunction, smoking, and chronic exposure to carbon monoxide. On examination of the skin, the finding of focal telangectasia is significant because of the association with pulmonary arterial venous fistuli, which cause arterial oxygen desaturation. Cardiopulmonary status should be carefully evaluated and the abdomen palpated for organomegaly. Acrocyanosis, pruritus, and splenomegaly have special implications in the diagnosis of polycythemia vera.

A classification of the polycythemia disorders is shown in Figure 54. It can be used as an algorithm for organizing the approach to diagnosis. The first question is whether the elevated hemoglobin level is caused by an actual increase in red cell mass or to a decrease in plasma volume, termed *relative polycythemia.* Common causes of relative polycythemia include a temporary decrease in plasma volume secondary to the use of diuretics or dehydration caused by nausea, vomiting, diarrhea, or excessive sweating. A sudden increase in vascular permeability, as with burns or anaphylaxis, also reduces plasma volume. Usually, the history identifies one of these mechanisms and dictates a logical approach to diagnosis.

Chronic plasma volume depletion in conjunction with a normal red cell mass occurs in Cushing's disease, with chronic catecholamine release and without apparent cause. The latter condition is referred to as *stress polycythemia* because of its frequent occurrence in hard-driving individuals with a history of smoking and cardiovascular disease, especially essential hypertension.[79,80] As a rule, these patients have modest increases in their hemoglobin and hematocrit (hematocrits of 50% to 55%). Definitive diagnosis as a patient with stress polycythemia requires a direct measurement of the red blood cell mass and plasma volume.[81] This is a

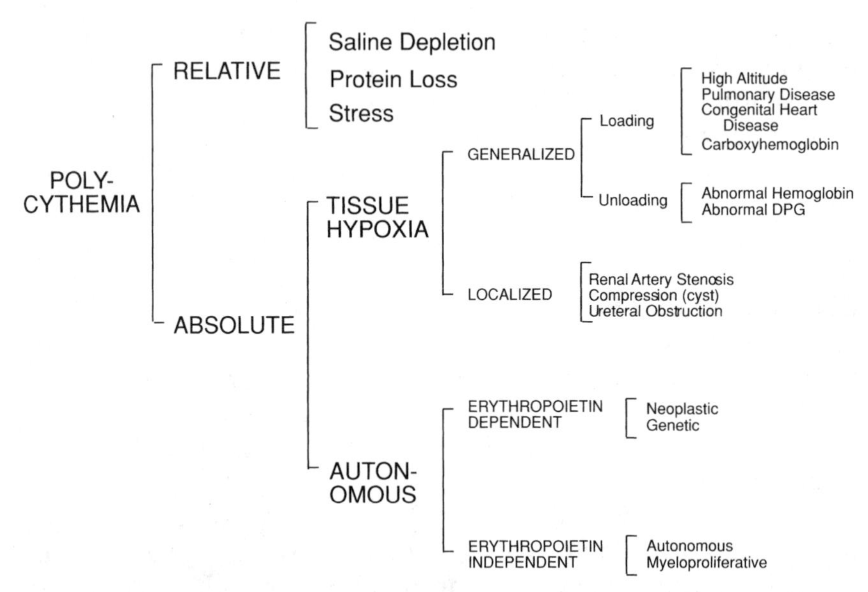

Figure 54. Classification of the polycythemic disorders. In the differential diagnosis of polycythemia, the first question to be asked is whether the elevated hemoglobin level is an absolute increase or a relative one secondary to plasma volume depletion. The absolute polycythemias may be subdivided on the basis of either a generalized or localized defect in tissue oxygen supply or the appearance of autonomous erythropoietin production or autonomous erythropoiesis. In the patient with hypoxia, studies of oxygen loading, unloading, and renal function can help identify the specific cause of the polycythemia. Patients with autonomous erythropoiesis, especially those with polycythemia vera, may be identified on the basis of their clinical presentation or detailed studies of erythropoietin secretion and stem cell growth characteristics. DPG=2,3 diphosphoglycerate.

technically difficult measurement. To improve the sensitivity of the measurement, it is important to carry out simultaneously a plasma volume study with radioalbumin and a red cell volume study using chromium-tagged red cells.

Hypoxia is a common cause of mild to moderate erythrocytosis. Unless the patient is severely cyanotic, the hemoglobin level rarely rises above 20 g/dL (Fig. 55). The diagnosis can be made by checking the arterial oxygen saturation. However, care should be taken to measure the carboxyhemoglobin level in any patient who is a heavy smoker or is at risk for environmental exposure to carbon monoxide. Patients who smoke two to three packs of cigarettes per day demonstrate carboxyhemoglobin levels of 10% to 15%,

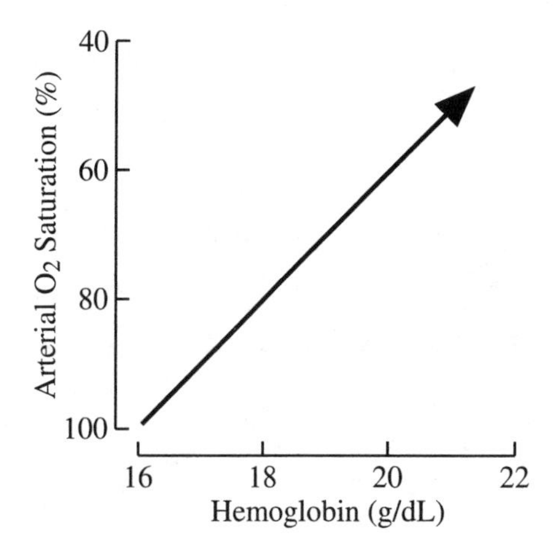

Figure 55. The relationship of hemoglobin to altitude hypoxia. There is a linear relationship between the hemoglobin level and the arterial oxygen saturation at altitude.

which by itself stimulates the hemoglobin to rise by 1 g/dL or more (hematocrit 50% to 55%).

The differential diagnosis of hypoxic polycythemia can be organized according to the alogrithm shown in Figure 54. Decreased oxygen loading of hemoglobin, as seen with pulmonary disease or with a right-to-left cardiac shunt, is easily detected by an arterial blood gas measurement. However, a single determination of arterial oxygen tension may not correlate with the degree of hypoxia experienced over 24 hours. In certain individuals, either sleep or physical activity may produce arterial desaturation that is not detected by a random sample.[82] Moreover, minute ventilation may be stimulated during the time of blood sampling, concealing an otherwise depressed oxygen saturation. Therefore, it may be necessary to carry out blood gas measurements on several occasions, both with exercise and during sleep.

The detection of an oxygen-unloading defect as a result of a hemoglobinopathy or a defect in 2, 3 diphosphoglycerate (DPG) production requires studies of the hemoglobin P_{50} and the oxygen dissociation curve (Fig. 56).[83] In about 50% these patients, a specific diagnosis can be made using hemoglobin electrophoresis.

Polycythemia can also result from localized renal hypoxia,[84] as demonstrated by experiments showing that partial obstruction of renal blood flow or an increase in intrarenal pressure through ureteral ligation will increase erythropoietin output and stimulate erythrocytosis. Clinically, similar effects are produced by unilateral renal disease from a variety of causes, including renal artery steno-

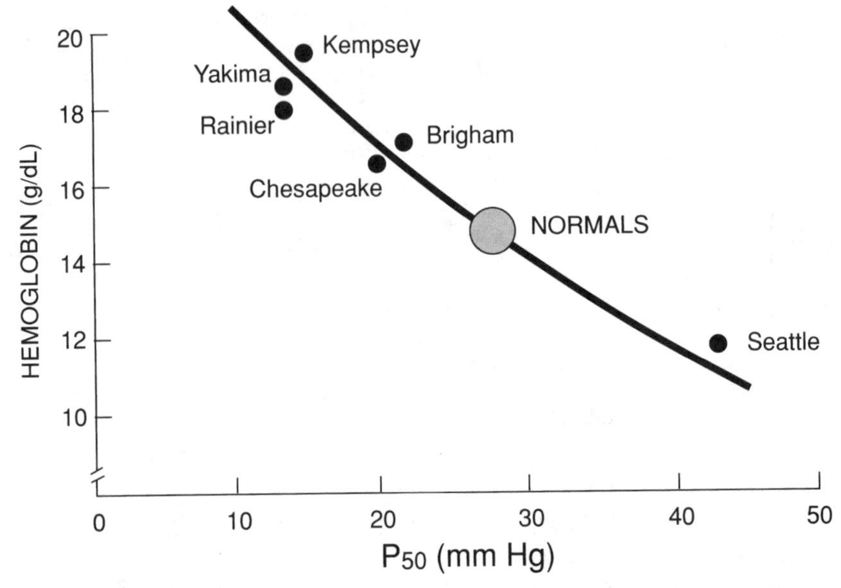

Figure 56. Hemoglobinopathies with abnormal patterns of hemoglobin oxygen dissociation. Inherited hemoglobin defects can result in abnormal oxygen unloading, characterized by a lower than normal P_{50}. These individuals compensate by modestly increasing their hemoglobin levels. In contrast, there are hemoglobinopathies where the P_{50} is increased and normal oxygen delivery occurs at a lower than normal hemoglobin level.

sis, cystic disease of the kidney, pyelonephritis, and ureteral obstruction. However, the renal lesion must be sufficient to produce local hypoxia but not enough to impair the function of the kidney's erythropoietin-producing cells. Consequently, the number of documented clinical cases is very small.

Polycythemia that is not the result of tissue hypoxia may or may not involve increased production of erythropoietin. Certain tumors, including hypernephroma of the kidney, angioblastoma of the cerebellum, and hepatomas, are associated with an increased output of an erythropoietin-like hormone.[85] The characteristic feature of this mechanism is the loss of feedback control of erythropoietin production. The autonomous nature of the erythropoietin production can be demonstrated clinically by serial measurements of serum erythropoietin levels before and after phlebotomy (Fig. 57).[86] In patients without demonstrable anatomic lesions, autonomous erythropoietin production and/or an in-

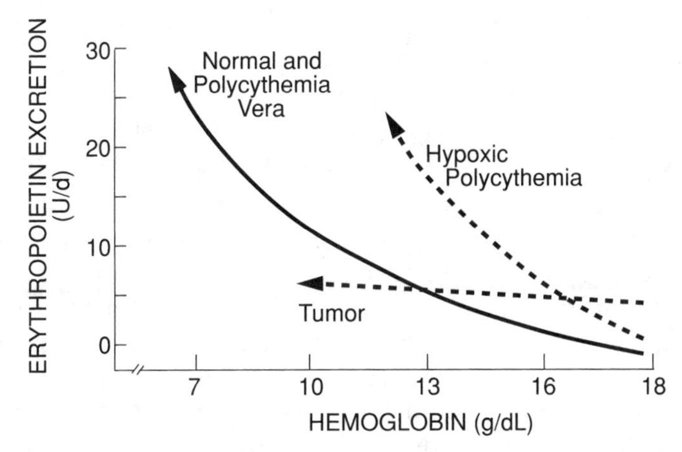

Figure 57. Patterns of erythropoietin excretion in patients with hypoxic polycythemia, polycythemia vera, and autonomous polycythemia secondary to tumor. Distinct patterns of erythropoietin excretion are observed when these patients are subjected to phlebotomy. In the case of polycythemia vera, the resultant increase in erythropoietin excretion is similar to that seen in normal patients. In contrast, the hypoxic polycythemia patient shows a more marked erythropoietin response to phlebotomy. The patient with autonomous erythropoiesis secondary to a tumor shows little or no response to phlebotomy, probably secondary to a suppression of normal secretion by the tumor hormone.

creased sensitivity to erythropoietin at the stem cell level (erythropoietin receptor and signal transduction protein defects) can occur as a familial disorder.[87]

Of all the polycythemias, polycythemia vera deserves particular comment. It is a relatively common acquired clonal malignancy of the stem cells of the marrow.[88] Although the malignant change affects all cell lines, its most prominent feature is a defect in the regulation of erythropoiesis. Red cell progenitors appear to be abnormally sensitive to erythropoietin, so increased proliferation occurs even though the in vivo erythropoietin level is suppressed. The pluripotent stem cell involvement is apparent from the presence of both leukocytosis and thrombocytosis in many of the patients, as well as from varying degrees of myelofibrosis and/or ineffective erythropoiesis. Over time, patients progress to a frank leukemic state resembling acute myelogenous leukemia.

Polycythemia vera can usually be diagnosed from its clinical manifestations. The peak incidence of the disorder is in the sixth and seventh decades of life. Patients present with complaints of headache, malaise, easy fatigability, and, very often, generalized pruritus. They may also suffer from gout caused by the high turnover of hematopoietic cells and overproduction of uric acid. On physical examination, over two thirds of the patients have splenomegaly. When the hemoglobin exceeds 20 g/dL, both plethora and acrocyanosis are usually observed. A routine CBC can be enough to confirm the diagnosis. A parallel increase in the numbers of red blood cells, granulocytes, and platelets is typical of polycythemia vera. To confirm the diagnosis, measurements of the number of circulating basophils and the leukocyte alkaline phosphatase level are commonly used. Most polycythemia vera patients demonstrate increases in both the basophil count and the leukocyte alkaline phosphatase. This pattern is distinctly different from the pattern seen with chronic myelogenous leukemia, in which the leukocyte alkaline phosphatase level is generally decreased.

The diagnosis of polycythemia vera is difficult if the increased proliferation is limited to the erythroid cells, if there is no palpable splenomegaly, or if the level of red cell proliferation is suppressed by iron deficiency or rendered ineffective by progressive myelofibrosis. In confusing situations, a measurement of the serum erythropoietin level may help. A low or undetectable serum erythropoietin level is suggestive of polycythemia vera. However, the renal mechanism for erythropoietin production is not impaired by the malignant change in the marrow stem cells. If, for any reason (myelofibrosis, blood loss, or iron deficiency), the hemoglobin level falls below the normal range, the serum erythropoietin level will show an appropriate increase (see Fig. 57). Therefore, a definitive diagnosis can be made only by demonstrating autonomous BFU–E or CFU–E growth on cell culture, a measurement available only in specialized research laboratories.[89]

Although the mechanisms leading to the various types of polycythemia are well understood, the differential diagnosis can be difficult. As suggested by the previous discussion, careful evaluation of the blood volume, arterial oxygen saturation, P_{50}, kidney structure and function, and, in some cases, the characteristics of stem cell growth may be needed. The separation of relative and absolute polycythemia based on the clinical presentation and blood volume measurements sounds simple, but the results can be quite misleading. Hypoxic polycythemia caused by decreased oxygen saturation may not be apparent from the clinical presentation or a single determination of the arterial blood gas. Hypoxia of one kidney or production of erythropoietin by a tumor can be very difficult to diagnose. Because the diagnosis of polycythemia vera is often made based on the exclusion of these other conditions, the clinician is at some risk of making a misdiagnosis. It is important, therefore, to observe the patient over time, monitoring the characteristics of the production of red blood cells, white blood cells, and platelets and their response to therapy.

Chapter

5 Management of Red Blood Cell Disorders

Successful management of a patient with anemia requires both an accurate diagnosis of the disorder and the skilled use of the best therapy, whether it is blood component transfusion, nutrient or vitamin replacement, growth factor administration, specific chemotherapy, or even bone marrow transplantation.

Transfusion Therapy

Modern blood banks collect and distribute a number of blood components for the treatment of anemia and coagulation disorders. The clinician is responsible for managing the patient's transfusion needs by assembling the appropriate mix of these components. For example, packed red blood cells compose the preferred transfusion product for correcting an anemia. Each "unit" of packed red cells contains approximately 200 mL of red cells suspended in 50 to 75 mL of plasma, giving a hematocrit of 70% to 80%. This transfusion unit is prepared from a donated unit of whole blood by gentle centrifugation. This fractionation procedure is also used to isolate the platelets and to prepare cryoprecipitate (fibrinogen and factor VIII concentrate) or fresh frozen plasma (FFP). As an alternative, single-donor platelets can be obtained by the technique of platelet pheresis. Blood centers also prepare red blood cell and platelet components that are filtered at the time of preparation to remove contaminating white blood cells. This step is becoming routine as a method to prevent alloimmunization of the recipient and cytomegalovirus (CMV) transmission.

Transfusion decisions must always be based on the disease process and the patient's ability to physiologically tolerate the anemia. Blood banks can minimize the risk of an adverse reaction by screening donors, testing the blood for viruses and antiviral antibodies prior to transfusion, and meticulously typing and crossmatching. However, transfusion therapy will never be risk-free, so red blood cell transfusions should be used only when absolutely necessary. The number of units transfused and the rate of transfusion must also be carefully supervised by the clinician according to the patient's condition. Blood centers publish guidelines to help prevent unnecessary transfusion and to guide clinicians in the management of patients who suffer a reaction. These guidelines need to be followed closely.

Acute Hemorrhage

The treatment of an acutely bleeding patient must begin immediately, even before transport to the hospital. Paramedical personnel are trained to control external bleeding and stabilize the patient in shock by initiating intravenous fluid therapy and applying military antishock trousers (MAST). When applied correctly, these trousers compress the venous system to produce a rapid redistribution of intravascular volume and provide an autotransfusion of 600 to 1500 mL of blood for perfusion of the heart, lungs, and brain. A MAST suit can also help to tamponade bleeding from arterial or venous vessels in the lower extremities of trauma patients.

The first and most urgent goal in treating patients with major hemorrhage is the rapid replacement of total blood volume.[90] Losses of up to 2 liters of whole blood in a patient who initially had a normal hemoglobin level is not enough to reduce the red cell mass to a critically low level. Any reduction in oxygen supply to tissues is more likely to be the result of poor perfusion. Therefore, treatment should concentrate on the rapid infusion of an effective volume expander (Table 29). A number of expanders are available, including noncolloidal (crystalloid or electrolyte) solutions, purified colloid preparations, and whole blood. Two standard crystalloid solutions, Ringer's lactate and normal saline, are immediately available and can be infused rapidly through intravenous needles or small catheters placed in peripheral veins. When given in volumes of 3 to 4 liters, an intravascular volume expansion of about 1 liter is accomplished because the electrolyte quickly equilibrates with the extravascular (extracellular) fluid space. This compartment is almost twice the size of the intravascular space, so a 3:1 ratio of infusate to intravascular volume expansion is the rule.

Table 29 • **Volume Expanders**

Crystalloid solutions:
Ringer's lactate
Normal saline
Colloid solutions:
5% albumin in normal saline
Plasma protein fraction preparation (PPF)
Hydroxyethyl starch (Hetastarch)
Whole blood (red blood cells plus fresh frozen plasma)

In younger patients with severe hemorrhage, electrolyte solutions can be given in volumes of up to 7 to 8 liters without complication. However, older patients are less able to tolerate these large volumes. In addition, patients with cardiac or hepatocellular disease have much more difficulty mobilizing excess salt and water from the extracellular fluid space. When there is renal damage, excessive electrolyte infusions can be life-threatening. Acute pulmonary edema and the more severe acute respiratory distress syndrome (ARDS or *shock lung*) are infrequent complications unless the patient has underlying cardiopulmonary disease, sepsis, or a crush injury.[91] When very large volumes of electrolyte solutions are given, it is not uncommon for peripheral edema to appear and for the patient to experience a reflex diuresis after normal intravascular volume status is restored.

When there is a sudden loss of more than 30% of the total blood volume, infusion of crystalloid can provide volume support for only the first 30 to 60 minutes, and some form of colloid solution should be used. Commonly available colloid solutions that provide reliable intravascular volume expansion include 5% albumin solution, plasma protein fraction (PPF), and hydroxyethyl starch (Hetastarch) solution. If available, 5% albumin solution is preferable. It gives reliable volume expansion without major side reactions and is hepatitis-free. Moreover, it is isotonic and stable when stored at room temperature for long periods. It can be infused as rapidly as crystalloid solutions, remains in the intravascular space, and, when bleeding has ceased, maintains adequate volume expansion for several days until the patient's own albumin production reconstitutes the normal plasma volume. In addition, as long as the volume infused is appropriate to the volume deficit, subcutaneous edema does not occur.

A common alternative to 5% albumin solution is PPF. PPF contains somewhat more than 4 g/dL of albumin, together with approximately 1 g/dL of immunoglobulins. These proteins are suspended in a sodium acetate solution that can vary from isotonic to hypertonic, with sodium levels as low as 110 to 120 mEq/L. In addition, the colloid level is quite low. If large volumes are used, the patient should be monitored for hyponatremia. Otherwise, the product is equivalent to 5% albumin solution. It is hepatitis-free and gives reliable volume expansion.

A stable solution of Hetastarch is available as a 6% solution in normal saline. This product also gives reliable intravascular volume expansion and has the advantage of being less expensive. Its therapeutic effectiveness is limited by its tendency to interfere with platelet function when given in large volumes (more than 2 liters).

In patients with severe hypovolemic shock, a crush injury, or sepsis, escape of protein and electrolyte into the lung can result in the development of ARDS. It has been suggested that crystalloid therapy is less of a threat to such patients. However, enthusiastic supporters of colloid replacement therapy argue that the increased intravascular osmotic pressure that results from colloid transfusion actually helps prevent fluid from entering lung tissue. This debate has not and probably will not be resolved. The best advice is to avoid overly aggressive fluid therapy of either kind in such patients.

Red Cell Replacement

Patients with severe blood loss also need red blood cell replacement. Young adults can tolerate acute reductions in their hemoglobin level to 50% of normal, approximately 7 g/dL, without compromising oxygen delivery to vital organs. However, older individuals or patients with cardiopulmonary problems need to be transfused earlier to a higher hemoglobin level. There is no simple guide to help the clinician determine how many units of red blood cells should be transfused. Helpful bedside observations include the amount of ongoing blood loss and both the cardiopulmonary and mental status of the patient. Frequent measurements of the hemoglobin or hematocrit can provide a rough guide to the need to transfuse, but only if volume expander therapy has been successful in restoring total blood volume. Total reliance on the hemoglobin and hematocrit measurements can be deceiving in patients who are experiencing a massive hemorrhage or who have not been treated aggressively enough with volume expanders.

Hospital blood banks maintain a supply of units of red blood cells of the most common blood types, including Type A, B, AB, and O. For each of these blood types, they maintain reserve supplies of both Rh positive and Rh negative blood. A red blood cell unit for a transfusion should be selected to be compatible with the recipient's ABO and Rh type. If time permits, the patient's serum should be screened for detectable antibodies using a pool of red cells that contain a range of red cell antigens. This procedure can take an hour or more, depending on whether there is evidence of a minor antibody that could result in an incompatibility. For patients who need emergency transfusion, type-specific blood should be used as a lifesaving procedure. Type-specific blood is ordered based on the determination of the patient's ABO type, a measurement that can be made within a few minutes. Generally, it can be transfused with little or no risk. Furthermore, the blood bank can complete the Rh typing and antibody screen measurement after the release of the blood for transfusion and, if serologic incompatibility is detected, warn the physician that red cell survival may be shorter than normal. Blood group incompatibilities outside of the ABO system only rarely result in intravascular red cell hemolysis.

If the patient's condition is desperate and red cell transfusion is necessary even before routine typing can be performed, type O, Rh negative red blood cells can be transfused. As suggested by the nomenclature, these red blood cells lack blood group A, B, and Rh substance on the cell membrane and are, therefore, antigenically silent. They can be transfused to patients of any blood type with little danger of intravascular hemolysis. At the same time, type O, Rh negative blood cannot be considered a transfusion panacea for the emergency situation. Only about 7% of donors are Type O, Rh negative, so the supply of this type of blood is always limited. The patient must be switched to type-specific blood as soon as the ABO type is determined.

Finally, in an emergency, Rh negative patients can be given transfusions of Rh positive red blood cells without fear of intravascular red blood cell hemolysis even when an anti-Rh antibody is present. In patients who do not have Rh antibodies, sensitization to Rh antigen following an Rh positive transfusion can be prevented by the administration of RhOgam (purified Rh antibody) within 1 or 2 days after transfusion.

In the average-size adult, a transfusion of 1 unit of red blood cells produces an incremental rise in the hemoglobin level of a little more than 1 g/dL, a hematocrit increase of 3% to 4% (Fig. 58). Serial measurements of the hematocrit following transfusion can be

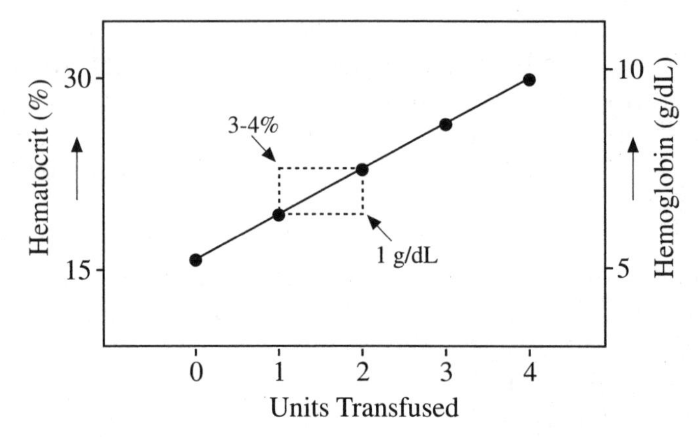

Figure 58. Response to transfusion. Each unit of packed red blood cells should result in a 1 g/dL rise in the hemoglobin level or a 3% to 4% rise in the hematocrit.

used, therefore, to guide the need for additional transfusions. When several units of packed red blood cells are given very quickly, the transfusion rate can be a problem. Even though only a small amount of plasma is infused with each unit, the combined effect of the red cells and plasma can represent a significant volume load for the patient. This must be taken into account in patients with marginal cardiopulmonary status. Rapid transfusion rates can also lead to more severe transfusion reactions. Patients can experience an immediate hypersensitivity reaction to vasoactive substances in the plasma or an IgA or IgE immune-mediated reaction to plasma proteins. With a severe reaction, the patient develops dyspnea, flushing, and urticaria with a loss of sufficient protein and electrolyte from the intravascular space to significantly reduce the plasma volume. Febrile reactions to cytokines in the transfused plasma and cellular HLA antigens can also be made worse by rapid transfusion.

From a practical standpoint, it is difficult to rapidly transfuse a unit of packed red blood cells. The high hematocrit of the red cell unit increases blood viscosity and slows its flow through the transfusion tubing and needle. Therefore, it may be necessary to dilute the unit by instillation of saline into the bag or by transfusion through a Y set wherein a simultaneous infusion of saline mixes with red cells before the flow through the narrow-gauge needle or intravenous catheter. Although this technique can speed the transfusion rate, it obviously adds to the problem in terms of the volume infused and the patient's cardiopulmonary status.

Coagulation Factor Replacement

With massive hemorrhage requiring multiple transfusions of colloid and packed red cells, significant falls in the patient's platelet count and levels of coagulation factors can occur. Electrolyte and purified colloid preparations lack coagulation factors, and unfiltered packed red blood cells stored at 4°C do not contain viable platelets. To maintain adequate hemostasis, any patient who has sustained a major injury needs a platelet count greater than 50,000/μL. If surgery is required, a count in excess of 100,000/μL may be necessary. The other coagulation factors are affected to a lesser degree. Levels of fibrinogen and factor VIII can actually rise in the face of acute hemorrhage or trauma because of increased release and/or production. As for key coagulation factors such as factors XI, X, IX, V, and prothrombin, levels below 30% are needed for abnormal hemostasis. These low levels are not easily attained. To dilute coagulation factors to levels below 30%, transfusion of more than 8 to 10 liters of factor-poor products is required.

The hemostatic defects associated with massive hemorrhage and transfusion are best managed by anticipating the problems. A platelet count, prothrombin time, partial thromboplastin time, and fibrinogen level should be routinely measured as a part of the initial workup of any bleeding patient. Platelet counts should then be repeated frequently to detect a significant fall and, if platelet transfusions are given, to document an adequate response. If there is concern regarding the presence of sepsis or an abnormal consumption of coagulation factors, a full coagulation profile should be obtained, including not only the platelet count, prothrombin time, partial thromboplastin time, and fibrinogen levels, but also a bleeding time, a thrombin time, and measurements for active fibrinolysis: fibrin split products and both antithrombin III and alpha-2 antiplasmin levels.

The first step in maintaining good hemostasis in massively transfused patients is to transfuse with enough platelets to increase the platelet count by 50,000 to 100,000/μL. This can be accomplished by giving a pack of random-donor platelets, in which 6 to 8 units of randomly donated platelets are pooled (Table 30). A unit of single-donor platelets generally provides a comparable result, that is, an incremental rise in the platelet count of 50,000/μL in an average-size adult. The need for FFP may also be anticipated based on the volume of blood loss or volume replacement and measurements of the prothrombin time and partial thromboplastin time.

Table 30 • **Coagulation Components**

Platelet Pack	
	From 6 to 8 U of pooled random donor platelets
Single-Donor Platelets	
	Equivalent to random donor "6-pack"
Fresh Frozen Plasma	
	Contains all coagulation factors at same concentration as normal plasma except for platelets
Cryoprecipitate	
	Contains factor VIII (antihemophilic factor) and fibrinogen
Prothrombin Complex Concentrates	
	Contains factors II, VIII, IX and is used in managing factor IX-deficient patients

Each unit of FFP contains approximately 200 mL of plasma with levels of coagulation factors approaching those found in fresh blood. Assuming a plasma volume of approximately 3 liters, each 200-mL unit will increase coagulation factors by about 5% to 7%. As a rule, a level of 30% or higher from the major coagulation factors should be sufficient for normal hemostasis. Therefore, an infusion of 4 to 6 units of FFP usually corrects the prothrombin and partial thromboplastin times sufficiently to provide adequate hemostasis as long as platelet supplementation has been adequate and no circulating anticoagulants (especially heparin) are present.

In patients with significant factor consumption because of disseminated intravascular coagulation (DIC) or fibrinolysis, platelet and FFP infusions alone may not be sufficient. In this situation, the cause of the DIC must be identified and adequately treated. After that, additional platelet transfusions and infusions of fibrinogen-rich cryoprecipitate can be used to reverse the thrombocytopenia and hypofibrinogenemia.

Chronic Anemia

A decision for transfusion in a patient with a chronic anemia always requires a careful weighing of benefits against risks. The purpose of transfusion should never be to normalize the hemoglobin level; it should be only to provide sufficient red blood cells to meet tissue oxygen needs. Most blood banks publish guidelines for the severity of anemia that deserves transfusion. The clinical evaluation of the patient's physiologic compensation to anemia is also extremely important. Any decision to transfuse must take into account the standard guidelines, the patient's physiologic status, and the natural course of the anemia, that is, whether the hemoglobin is expected to decrease further or to increase based on a compensatory erythropoietic response.

The ability to compensate for a chronic anemia varies considerably with age and physical condition. A young person not involved in demanding physical work can tolerate hemoglobin levels of 6 to 8 g/dL without cardiovascular compromise. In contrast, elderly individuals can experience easy fatigability, claudication with exercise, angina, and congestive heart failure even when the anemia is more moderate. The overall impact of severe anemia on health status is best illustrated by experience in treating patients having renal-failure with erythropoietin. General well-being and exercise physiology improve steadily as the hemoglobin increases to levels of 10 to 11 g/dL. Above that, there appears to be little further improvement.

The complications of red blood cell transfusion, although infrequent, are to be avoided whenever possible. They include acute volume overload, immunosensitization, bacterial or viral infection, transfusion error, and, with multiple transfusions, iron overload. Transfusion in an individual with a chronic anemia is quite different from that in an actively bleeding patient. Patients with chronic anemia often have a larger-than-normal total blood volume. If several units of packed red blood cells are administered in a short period (less than 24 hours), blood volume and blood viscosity increase, raising the threat of circulatory overload and acute cardiac failure. In this situation, it should be standard practice to reduce the patient's blood volume by the use of a diuretic or a limited exchange transfusion. The latter technique involves phlebotomizing the patient of 200 to 400 mL of whole blood immediately before administering 2 to 3 units of packed red blood cells.

The risk of transfusion hepatitis has been significantly reduced by pretransfusion screening, but the transmission of non-A, non-B hepatitis (hepatitis C) cannot be completely prevented. The CMV status of both the patient and the transfused blood should also be determined. Patients who are CMV-negative and severely immunocompromised (as with bone marrow or renal transplantation) should receive CMV-negative blood. For patients at less risk, well-filtered packed red blood cells appear to be adequate to prevent CMV infection.

Any ABO-incompatible blood transfusion carries a risk of intravascular hemolysis and renal failure. Patients should be monitored closely during transfusion for signs of reaction. Any complaint of discomfort, especially low back pain, should alert the staff to a possible intravascular hemolytic transfusion reaction. The transfusion should be stopped immediately and a blood sample drawn, immediately centrifuged, and the plasma inspected for intravascular hemolysis. Renal damage can be prevented if the transfusion is halted quickly enough.

The development of a minor blood group antibody is not associated with intravascular hemolysis. It is usually detected by the antibody screening test (DAT) at the time of the next transfusion. Future reactions can be prevented by selecting donors who lack the antigen specific for the recipient's antibodies. However, when the patient has several antibodies, finding compatible blood may be extremely difficult.

Because neither white blood cells nor platelets can be routinely matched to the recipient, transfusion of multiple units of red blood cells containing contaminating granulocytes and platelets can lead to human lymphocyte antigen (HLA) alloimmunization.[92] The sensitized patient is recognized clinically by the appearance of rigors and febrile reactions during or immediately following a red blood cell or platelet transfusion. Sensitization can be prevented, or at least ameliorated, by filtering the red blood cell unit at the time of either donation or administration.

Finally, it is important to recognize the limitations of transfusion therapy in specific disease states. For example, hypoproliferative anemias caused by inflammatory states generally are relatively mild and should not be treated with transfusion.The anemia is largely physiologic and does not cause symptoms. Furthermore, it disappears if and when the inflammatory state resolves. The hemoglobin level rarely falls below 10 g/dL unless the condition is especially severe or there is a complicating illness, such as chronic blood loss with iron deficiency or severe malnutrition with protein depletion. If such patients receive transfusions, the response is transitory. The hemoglobin quickly returns to the original level because marrow production is insufficient to sustain the artificially increased red cell mass. Thus, transfusion should be used only when clearly needed, for example, for protection during a surgical procedure. More severe anemia in patients with renal disease should be treated with erythropoietin. In most of these patients, this therapy eliminates the need for chronic transfusion. This is much more physiologic and prevents the complications of transfusion, especially the possibility of iron overload.

Growth Factor Therapy

A number of hematopoietic growth factors, including EPO, G-CSF, GM-CSF, and thrombopoietin (TPO), have been isolated and (using recombinant technology) produced as therapeutic agents. EPO was the first growth factor to be applied clinically using a recombinant human erythropoietin (rHuEPO) produced commercially in a mammalian cell system. The pharmaceutical preparation of EPO is virtually identical to human EPO.

The therapeutic applications of EPO are now well defined. The first and most obvious use is in patients with severe renal disease.[93,94] In this situation, in which endogenous production of EPO is severely compromised, relatively low-dose therapy (25 to 150 μ/kg, intravenously or subcutaneously, three times a week) eradicates the transfusion requirement and over a period of weeks raises the hemoglobin to levels of 11 to 12 g/dL. The correction of the anemia is accompanied by an improvement in both exercise physiology and general well-being, although increases above the 11 g/dL hemoglobin level do not appear to have added benefit and may be a risk to the patient. Aggressive EPO therapy can aggravate the patient's hypertension, secondary to an increase in blood volume. Dialysis patients may also show an increased tendency to fistula clotting as the anemia is corrected.

The key to successful EPO therapy is the maintenance of an adequate iron supply to the marrow. All patients should be carefully evaluated before therapy to assess iron status. Patients with serum ferritin levels below 150 μg/L are at risk for a delayed or substandard response to EPO therapy. They need aggressive iron supplementation, either with daily oral iron or periodic injections of intravenous iron dextran. Furthermore, iron deficiency can appear at any time during the patient's maintenance therapy, requiring repeated assessments of iron status and compliance with iron therapy.

Inflammatory cytokines can also suppress the erythroid marrow's response to EPO, so that, during periods of acute illness, it may be necessary to supplement the patient's EPO therapy with transfusion. Other conditions that can result in EPO resistance include aluminum toxicity and secondary hyperparathyroidism.

EPO therapy has been approved for a number of other clinical conditions. In AIDS, marrow suppression can be a treatment limiting toxicity for patients receiving zidovudine (AZT). EPO therapy counteracts this effect and allows patients to continue on AZT treatment.[95] An assessment of the plasma EPO level can help guide therapy. Patients who do respond generally have plasma EPO levels lower than 500 IU/L. When the plasma EPO level is higher than that, little or no effect is observed using standard doses of rHuEPO.

Other clinical applications for the hormone include treatment of the anemia of prematurity and the prevention of anemia in patients receiving chemotherapy. EPO has been used preoperatively to stimulate red blood cell production for autologous blood storage and perioperatively to encourage anemia repair.[96] It can be effective in both settings, although the cost of therapy can be much higher than for standard transfusion practice. The use of rHuEPO in such disorders as the anemia of chronic disease or in patients with myelodysplasia, aplastic anemia, and myelofibrosis is less clear. In general, anemias secondary to an inflammatory disorder are relatively mild and not clinically symptomatic. However, when they are severe, inflammatory anemias (the anemia of chronic disease) can be successfully treated with rHuEPO. When the hormone is given in therapeutic amounts, it overcomes the cytokine-induced block in iron supply. Patients show a predictable response to therapy, although it may be somewhat delayed. As for patients with marrow-damage anemias, endogenous EPO levels tend to be high, and successful therapy with standard doses of rHuEPO is unlikely.

Both G-CSF and GM-CSF are now used extensively in the treatment of neutropenic patients, especially those recovering from chemotherapy-induced granulocytopenia. These hormones have been extremely valuable in speeding the recovery from neutropenia in patients undergoing bone marrow transplantation or ablative chemotherapy. Each growth factor is amazingly lineage-specific. G-CSF and GM-CSF have virtually no clinical effect on red blood cell production, just as EPO does not stimulate the proliferation of white blood cell precursors. The growth factor responsible for platelet production, TPO, may show cross-reactivity with EPO. The hormones are quite similar in structure and thus may bind with both EPO and TPO cell receptors. This cross-reactivity needs to be better defined as clinical testing of TPO goes forward.

Treatment of Nutritional Anemias

The term *nutritional* is used to cover a group of anemias that result from a single or combined deficiency of iron, folic acid, or vitamin B_{12}. There are a number of other unusual nutritional anemias including pyridoxine (Vitamin B_6), copper, and riboflavin deficiency. Protein malnutrition, per se, can affect body metabolism and produce a mild anemia that is corrected by protein refeeding. This anemia represents an adjustment to a lowered body metabolism and by itself is of little importance.

Although the principal agenda in the treatment of any nutritional anemia is the replacement of the key nutrient, successful therapy may depend on the anemia's underlying cause, whether it is dietary deficiency, increased metabolic requirement, or excessive loss. Therefore, a careful evaluation of the patient to define the basis of the deficiency state is essential and dictates management. For example, the treatment of a patient with iron deficiency secondary to malabsorption may not respond to oral iron therapy and needs to be treated with intravenous iron. Diseases of the small intestine can also result in simultaneous deficiencies of several nutrients. This may not be recognized initially because one of the deficiencies is usually dominant. However, the others emerge as the patient is treated, requiring combination therapy.

Iron Deficiency

The most common nutritional anemia is caused by iron deficiency. Iron absorption and iron storage are highly regulated in humans. As iron stores decrease, iron absorption compensates by increasing the percentage of iron absorbed from food or medicinal iron. Patients with iron-deficient erythropoiesis or iron-deficiency anemia can demonstrate a fourfold to fivefold increase in absorption. At the same time, intestinal mucosal cells will not absorb unlimited amounts of iron. Only a portion of medicinal iron given orally is absorbed, and the percent absorption decreases as the iron dose is increased (Fig. 59). At the recommended oral iron dose of 200 mg per day given in three equally spaced doses between meals, about 40 to 60 mg is absorbed and delivered to the erythroid marrow. This amount is sufficient to support a red blood cell production rate of two to three times normal, for an increase in hemoglobin of 0.2 g/dL per day. If a higher oral dose is given, the amount absorbed does not progressively increase, and gastrointestinal side reactions become much more pronounced. This is a reflection of the intesti-

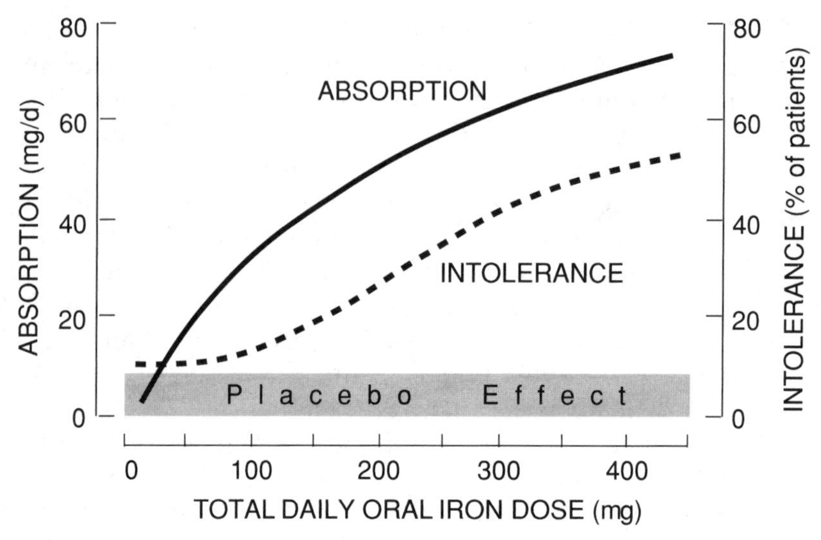

Figure 59. Iron absorption and tolerance. With increasing doses of oral iron, there is a decrease in the amount absorbed, whereas the incidence of gastrointestinal side effects increases. For this reason, it is recommended that most patients be treated with 200 mg of iron per day given in at least three equally spaced doses between meals. This will result in the absorption of 40 to 60 mg, with less than one third of patients demonstrating significant side reactions.

nal mucosal cells' ability to set an upper limit for iron absorption, regardless of the patient's iron deficiency state.

When prescribing oral iron, the important rules are to administer a ferrous iron salt (ferrous sulfate, fumarate, or gluconate tablets, or ferrous sulfate syrup) and to prescribe the dosage according to the elemental iron content. Most iron tablets contain 60 to 70 mg of elemental iron, so that a daily schedule of 3 to 4 tablets provides the patient with at least 200 mg of elemental iron each day. In children, the liquid preparation is easier to administer, and for children weighing less than 35 kg, a total dose of 50 mg per day is sufficient. Iron absorption is improved by giving the iron preparations between meals. It is also essential that the tablet dissolve rapidly in the stomach. Gastrectomized patients with low stomach acidity and rapid food transit times may be unable to dissolve the tablet coating. In this case, ferrous sulfate syrup should be used. Preparations are sold that contain additives such as vitamin C or small amounts of amino acids to enhance iron absorption. They are generally much more expensive, but not more effective.

The amount of iron given should be determined by the rate of response required, by considerations of gastric tolerance, and by evidence of continued blood loss. If the anemia is nutritional and of minimal severity (e.g., the iron deficiency that develops during pregnancy), a multivitamin preparation containing iron or a single pill containing 60 to 70 mg of elemental iron taken once a day is generally sufficient to re-establish normal iron balance. For more severe anemia, doses up to 200 mg per day are usually well tolerated, although the incidence of side reactions begins to increase (see Fig. 59). Attempts to give larger doses will result in gastrointestinal intolerance, particularly heartburn, nausea and gastric discomfort in up to 50% of patients. Constipation and diarrhea appear less related to dose. To control side reactions, the daily dose may be reduced or the iron preparation may be given with meals to encourage compliance. Although the decrease in iron intake prolongs the recovery phase, oral iron therapy is preferable to such alternatives as parenteral iron or transfusion.

The response to oral iron is best followed by monitoring the hemoglobin level over a period of 2 to 3 weeks. A rise in the reticulocyte count during the second week can also indicate a response, although it is less precise and less quantitative. A failure to respond can reflect continued blood loss, malabsorption of iron, or an incorrect diagnosis, as for a patient with the iron-deficient erythropoiesis of inflammation. When a patient demonstrates severe iron malabsorption, iron can be administered parenterally in the form of iron dextran injection (InFeD). This substance is a complex of ferric hydroxide and dextran. It is packaged as a colloidal solution containing 50 mg/mL of iron in 2 and 5 mL ampules for intramuscular or intravenous administration. The preparation is best given intravenously. Considerable caution must be exercised in using intravenous iron dextran, since it can be associated with a severe, even fatal, anaphylactoid reaction. All patients must be first tested by injection of a few drops of the material followed by close observation for 5 to 10 minutes for hypotension or complaints of headache, back pain, chest pain, or anxiety. An infusion of 500 mg or more may then be given over a period of 10 to 30 minutes with continued observation. The patient's blood pressure should be monitored at frequent intervals and the infusion discontinued immediately if the patient becomes symptomatic or the blood pressure falls.

The iron dextran material is removed from the plasma by the reticuloendothelial cells, which process it, releasing the iron for return to the plasma and erythroid marrow. The rate at which the iron becomes available does not exceed that which can be achieved by oral iron therapy when absorption is normal. Therefore, iron dextran should not be used simply to speed the rate of recovery. Late reactions to iron dextran sometimes occur, including a serum sickness-like condition characterized by fever, lymphadenopathy, arthralgias, and urticaria. The frequency of late reactions is higher in patients who suffer from rheumatoid arthritis. The use of iron dextran should be restricted, therefore, to situations in which effective oral iron therapy is clearly impossible.

Vitamin B_{12} Deficiency

Vitamin B_{12} deficiency may be anticipated in patients with gastrointestinal defects or may be recognized from its characteristic hematopoietic and nervous system manifestations. The diagnosis is usually readily made from the combination of a megaloblastic anemia, a low serum vitamin B_{12} level, and, if necessary, a measurement of vitamin B_{12} absorption. It is important, however, to look for simultaneous deficiencies of other nutrients including folic acid and iron. Patients with malabsorption frequently demonstrate combined deficiencies. Moreover, the response to vitamin B_{12} therapy is dependent on the adequacy of both folic acid and iron supply.

Vitamin B_{12} is available as cyanocobalamin injection, USP, in concentrations of 30, 100, and 1000 μg/mL in vials containing 1 to 30 mL. It is administered in doses of 1 to 1000 μg by either intramuscular or deep subcutaneous injection. Effective transport and storage of the injected vitamin B_{12} depend on the binding capacity of the transcobalamins. Most of an individual dose in excess of 100 μg is rapidly excreted in the urine. Therefore, to initially treat a vitamin B_{12} deficient patient and replenish the liver vitamin B_{12} stores, it is best to administer 100 μg per day for several weeks rather than to administer larger, less frequent doses.

Oral vitamin B_{12} preparations are available for the treatment of pure nutritional deficiency states. Preparations that combine vitamin B_{12} with intrinsic factor are also available but are not considered reliable. In some patients, antibodies against the intrinsic factor-B_{12} complex may prevent absorption from the outset; in others, refractoriness develops with constant use. Oral B_{12} therapy with very large doses has been used in some patients for maintenance therapy.[97] However, to guarantee a sustained remission, parenteral administration is preferred and needs to be continued for life. It is essential, therefore, to have good patient compliance and to develop a reliable system to guarantee the parenteral administration of at least 100 μg of cyanocobalamin every 2 to 4 weeks.

The pattern of response to vitamin B_{12} therapy is shown in Figure 60. The megaloblastic marrow morphology begins to disappear within hours, and the plasma iron and bilirubin concentrations fall rapidly as the ineffective erythropoiesis is corrected. An increase in the reticulocyte index is observed on the second or third day and reaches a peak 3 to 5 days later.[98] Following these early changes, the hemoglobin level begins to rise at a rate of 0.3 to 0.6 g/dL per day. The rate of recovery depends on several factors, including the severity of the anemia, the adequacy of iron stores, renal function, and the patient's general health. If, after an early response, the reticulocyte production index declines and the rise in hemoglobin is curtailed, a complicating inflammatory illness or lack of adequate iron stores should be suspected. In patients with pancytopenia, the platelet count generally rises rapidly and briefly overshoots the normal level. The granulocyte count recovers more slowly. Al-

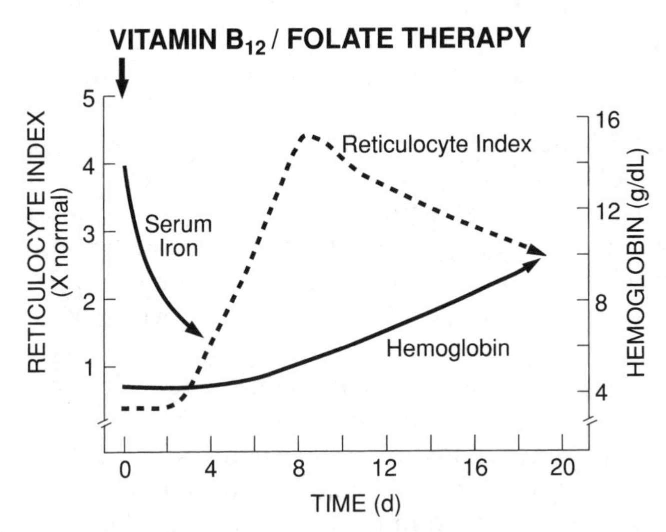

Figure 60. Response to vitamin B_{12} folic acid therapy. Within hours of initiation of appropriate therapy, the megaloblastosis begins to disappear and the serum iron falls. An increase in the reticulocyte index is first observed on the second to third day and peaks by the eighth day. It may then decline, depending on the level of iron supply. The hemoglobin level responds more slowly, rising at a rate of 0.3 to 0.6 g/dL per day. A full correction of the anemia can take several weeks.

though the patient usually experiences an increased sense of well-being within the first 24 hours of therapy, well-established neurologic defects such as ataxia or loss of position and vibratory sense may be fixed or improve only slowly over 6 months or more of therapy.

Folic Acid Deficiency

Folic acid deficiency is most commonly seen in patients with diseases of the small intestine and in chronic alcoholics. The megaloblastic anemia of folate deficiency cannot be distinguished from that of vitamin B_{12} deficiency, but measurements of serum and red cell folate levels make accurate diagnosis possible. Moreover, folate deficiency is rarely associated with the specific neurologic abnormalities that are characteristic of vitamin B_{12} deficiency. If the etiology is in question, combination therapy with both folic acid and vitamin B_{12} should be given. The diagnosis can then be made after treatment is begun, from analysis of vitamin levels obtained before treatment and studies of vitamin B_{12} absorption.

Folic acid, USP (folvite), is marketed as 0.1-, 0.4-, 0.8-, and 1-mg tablets of pteroylglutamic acid. It is also available in combination with a number of other vitamins and/or iron in multivitamin preparations that contain from 100 μg to 1 mg of the vitamin. Folic acid injection is available in the concentration of 5 mg per mL in a 10-mL multidose vial.

Excessive alcohol intake is associated both with a poor dietary intake of folate and a block of the release of folate from stores. Therefore, management requires both folate administration and alcohol withdrawal. Patients with primary malabsorption disorders frequently have combined folate and iron deficiency, which can interfere with the hematologic response. Even with severe malabsorption, the folate deficiency can usually be treated with an oral dose of 1 mg of folate twice daily. Parenteral folate therapy is, therefore, usually reserved for the patient who is taking nothing by mouth. Larger doses of folate can be given without risk of toxicity if there is concern regarding the patient's ability to absorb the vitamin. However, folic acid in large amounts may increase the frequency of seizures in children taking anti-epileptic medications.

If a patient with vitamin B_{12} deficiency is given large doses of folic acid by mistake, an improvement in the megaloblastic anemia may be observed as the folate circumvents the methyl folate trap.

At the same time, such therapy does not prevent or alleviate the neurologic defects of vitamin B_{12} deficiency, which may then become irreversible. This also applies to the prophylactic administration of folic acid, which should always be reserved for situations in which there is a clear risk of folate deficiency but no threat of vitamin B_{12} deficiency. Two common situations that fit this description include pregnancy and any of the severe hemolytic anemias. If folic acid is to be given to a patient as a therapeutic trial, the daily dose should be less than 100 μg, given intramuscularly. This will produce a hematologic response in the patient with pure folate deficiency, but not in the patient with vitamin B_{12} deficiency.

Miscellaneous Deficiency States

Copper deficiency can be observed in patients on total parenteral nutrition (TPN) and after intestinal bypass surgery. Malnourished infants and infants with a copper-deficient diet may also develop granulocytopenia and anemia. A daily dose of 0.1 mg/kg of copper sulfate given by mouth, or up to 0.05 mg/kg added to the daily TPN formula, is sufficient to correct the abnormality.

In patients receiving anti-tuberculous drugs, (isoniazid and/or pyrizinamide), pyridoxine deficiency can be prevented with a dose of 50 mg of pyridoxine by mouth each day. Rarely, in sideroblastic anemias of unknown cause, giving oral pyridoxine therapy of 50 to 100 mg a day for prolonged periods produces a partial response. A few case reports have also suggested that some sideroblastic anemias respond to vitamin C, crude liver extract, and L-tryptophan. In alcoholic patients with ring sideroblasts, folic acid deficiency appears to play a major role. In this circumstance, withdrawal from alcohol and treatment with folic acid is usually sufficient to correct the sideroblastic defect, although simultaneous treatment with oral pyridoxine is not unreasonable.

Immunosuppressive Therapy

Two major categories of immune disease result in anemia: (1) an immune suppression of stem cell proliferation and maturation resulting in selective red cell aplasia or aplastic anemia, and (2) the production of antibodies that act on circulating red cells to produce a hemolytic anemia. Because they are different processes, they require different therapeutic approaches.

Stem Cell Disease

The sudden appearance of pure red cell aplasia or an aplastic anemia is often caused by an immune process, either antibodies directed against the stem cell or a cellular immune mechanism (T lymphocytes) that suppresses stem cell proliferation. The latter may also occur transiently with certain viral infections, such as parvovirus, or as a result of drug action. Pure red cell aplasia is often associated with thymomas. Some patients with red cell aplasia have been shown by cell culture techniques to exhibit either a humoral agent or T cell suppression that responds to high-dose steroids in vitro, thereby providing a guide to therapy. In this situation, high-dose steroids and/or immunosuppressive therapy with cyclosphosphamide, antithymocyte globulin (ATG), or antithoracic duct lymphocytic globulin (ATDLG) may be effective.

Aplastic anemia (i.e., total marrow failure with a reduction in circulating levels of red cells, granulocytes, and platelets) can also result from an immune process. It is always important to rule out a malignant process, such as acute leukemia, where specific chemotherapy is indicated. For patients with aplastic anemia, whose marrow is virtually devoid of hematopoietic cells, immediate consideration must be given to either ATG therapy or bone marrow transplantation.[99] The severity of the pancytopenia and the relative cellularity of the marrow biopsy can be used to guide the course of action. If the circulating granulocyte is less than 500/μL, the platelet count less than 20,000/μL, the reticulocyte count less than 1%, and marrow cellularity less than 25% of normal, the prognosis is extremely poor. The marrow is unlikely to recover spontaneously, and most patients will die of infection or bleeding complications within 1 to 2 years. This does not mean that a reversible cause should be ignored. As a part of the evaluation of any patient with aplastic anemia, an exhaustive search for a reversible cause should be undertaken. Any drug that might be implicated should be discontinued immediately and blood and marrow examinations repeated over several days to help identify rapidly reversible lesions.

As with pure red cell aplasia, immunosuppressive therapy, particularly with ATG and ATDLG, has been shown to be effective in more than one half of patients with aplastic anemia.[100] This form of therapy is preferred in patients over 40 years of age because of the increased morbidity and mortality of marrow transplantation in older individuals. Unfortunately, the success rate seems to depend on the source of the ATG or ATDLG and the ability of the patient to tolerate a full course of therapy. Androgen therapy has been used in the past to treat such patients, but with relatively little success.

Bone marrow transplantation for aplastic anemia is recommended in younger patients who have an identical twin donor or an HLA-matched donor.[101] In patients given transplants early, survivals of greater than 2 years are reportedly 60% to 70%. Transplantation may be done in identical twins without preparation simply by infusion of adequate numbers of hematopoietic stem cells from the normal twin. In the case of the HLA-matched marrow transplant, the recipient must be pretreated with intensive immunosuppressive conditioning therapy, including both high-dose cyclophosphamide and total body irradiation. This treatment effectively suppresses the recipient's immune system, thus preventing rejection of donor stem cells. It may also encourage a spontaneous remission by interfering with an autoimmune process.

Allogenic bone marrow transplantation is not without risk. As many as 5% to 15% of patients do not survive the transplant procedure. The treatment also increases the patient's susceptibility to infections. Approximately 35% of HLA-matched transplant recipients develop significant graft-versus-host disease (GVHD), characterized by skin rash, hepatitis, and enterocolitis. The basis for the GVHD is the transfusion of immunocompetent T cells, which react against recipient tissues. The severity of GVHD can be blunted by post-transplant treatment with methotrexate and cyclosporin. Severe GVHD can result in the early death of the patient during the recovery period; chronic GVHD can develop some 6 to 12 months after the transplant and, if severe, also shorten the patient's life span.

Even in the best circumstances, bone marrow transplantation is not the therapeutic answer for most patients. Despite the use of both matched sibling donors and a nationwide pool of nonrelated matched donors, less than one half of patients can receive transplants. Transplant rejection and early death are risks and are more common in patients who have received multiple transfusions before the procedure. Age is another major factor. Survival is best in young patients. As recipients become older, the complications, especially severe GVHD, increase. Based on this knowledge, patients under 40 years of age with severe aplastic anemia should receive transplants without delay if a donor is available. Older patients and patients who have received multiple transfusions should be treated with aggressive immunosuppressive therapy.

Immune Hemolytic Anemia

The management of an autoimmune hemolytic anemia depends on the underlying disease process and an accurate definition of the antibody involved, the rate of red blood cell destruction, and whether the hemolysis is intravascular or extravascular.[102] Severe, acute intravascular hemolysis from a transfusion of incompatible red blood cells is a life-threatening event. The release of free hemoglobin and cell membrane fragments may produce renal failure. In this situation, the transfusion must be discontinued immediately and mannitol administered to stimulate a brisk diuresis. Generally, the renal failure associated with a mismatched transfusion is reversible.

In the treatment of a patient with an intravascular hemolytic anemia secondary to an autoantibody, the success of a red blood cell transfusion depends on the antibody titer and the level of complement. In patients whose plasma contains high-titer IgM antibodies, the rate of hemolysis may be limited by a complement depletion. In this situation, one should avoid transfusion of even small amounts of plasma-containing complement, even the amount present in the plasma of a unit of packed red blood cells. These patients should be transfused only with washed red blood cells, from which all traces of plasma have been removed. The survival of washed red blood cells is generally no better or worse than that of the patient's own red cells. Over time, however, patients with cold agglutinin (IgM) hemolytic disease demonstrate an increased transfusion requirement, as the titer of the IgM antibody and/or temperature specificity increases.

Cold-agglutinin hemolytic disease is usually unresponsive to steroid therapy. Greater success may be anticipated with the use of alkylating agents such as chlorambucil and cyclophosphamide. This is particularly true when a lymphoproliferative disorder is involved. In addition, splenectomy is less likely to provide significant benefit. For patients in whom the temperature specificity of the antibody approaches room temperature, cold exposure can result in acute intravascular hemolytic events. Therefore, particular attention must be given to protecting such patients from cold, especially chilling of the extremities.

Patients with a warm-reacting IgG antibody with or without associated complement binding are more responsive to treatment. Red blood cell destruction is primarily extravascular, so there is little risk of renal damage. Furthermore, red blood cell transfusions can be given without the risk of an acute hemolytic event because the rate of hemolysis is not dependent on complement levels. Most of these patients are very responsive to steroid therapy. When prednisone is given in a dose of 1 mg/kg per day, red blood cell destruction is significantly reduced within days or a few weeks. The response to treatment is most easily followed by observing the reticulocyte count and hemoglobin concentration. The early response may reflect the ability of steroids to interfere with the reticuloendothelial cell destruction of the antibody-coated red blood cells. However, the more important action of steroids is to decrease the production of the autoantibody by way of its lympholytic effect. This can be monitored by serial measurements of the patient's direct antiglobulin test (DAT or Coombs test). As the antibody titer falls, the DAT becomes weaker and in some patients reverts to normal. Splenectomy is recommended for patients whose anemia cannot be controlled with steroids or a combination of steroids and an alkylating agent.

In patients with lymphoproliferative malignancies and autoimmune hemolytic anemia (AIHA), chemotherapy can usually control the hemolytic process. Intravenous immunoglobulin and plasma pheresis have been used with some success in AIHA patients. Extracorporeal absorption of IgG with an anti-staphylococcal protein-A-silica column may help in refractory patients with life-threatening disease. Red blood cell transfusions should be avoided unless the patient's anemia is symptomatic. In that case, red blood cells should be given slowly and the patient's response closely monitored.

Other Forms of Therapy

Management of Hypoproliferative Anemias

Most hypoproliferative anemias are relatively mild and represent manifestations of a systemic disease process. Therefore, they usually do not require treatment. For example, the hypoproliferative anemia seen with hypothyroidism or panhypopituitarism reflects the overall decrease of tissue oxygen requirements. The severity of the anemia is appropriate for the reduction in the patient's basal metabolism. Furthermore, the anemia is automatically corrected once hormonal supplementation is initiated. Similarly, the anemia of protein malnutrition responds to refeeding as the patient's metabolic status improves. Transfusion in these situations is appropriate only when the patient is elderly or demonstrates severe cardiovascular disease. Specific nutrients such as iron, folic acid, and vitamin B_{12} are indicated if, because of malnutrition or malabsorption, the patient is at risk for a nutritional anemia.

The treatment of any inflammatory anemia should always be directed at the disease responsible for the anemia. For example, the hypoproliferative anemia of patients with rheumatoid arthritis may be reversed with effective anti-inflammatory therapy. When a patient does not respond, it is important to look for exacerbations of rheumatoid activity or a complicating factor such as iron deficiency or renal failure. Chronic blood loss associated with long-term treatment with steroids or nonsteroidal anti-inflammatory drugs is a frequent complication. Therefore, if an arthritis patient demonstrates worsening anemia and iron studies suggest iron store depletion, a therapeutic trial with oral iron is in order. Rheumatoid arthritis patients can be effectively treated with EPO if their anemia is especially severe and not responsive to anti-inflammatory therapy. Recombinant EPO in doses of 150 units/kg three times a week will, over a period of weeks, produce a sustained rise in the patient's hemoglobin level. This demonstrates the ability of EPO to overcome the cytokine impact on stem cell growth and iron delivery. Patients treated with EPO should be maintained on oral iron and carefully monitored for the development of iron deficiency.

Hydroxyurea in Sickle Cell Anemia Patients

After a series of clinical trials, hydroxyurea has been approved for the treatment of sickle cell anemia.[103] Maintenance therapy with doses of 10 to 30 mg/kg per day increases the red blood cell Hb F level in most patients. Patients with relatively high baseline Hb F levels show the best responses. Hb F levels can rise 20% to 30%. At the same time, both the MCV and total hemoglobin concentration will rise, matched by a decrease in the absolute reticulocyte count. This change is beneficial to the overall clinical picture, including a significant reduction in the number of painful sickle crises.[104] Other cell-cycle–specific agents, such as 5-azacytidine and Ara-C, have also been shown to increase Hb F levels.

Splenectomy in Hemolytic Anemia Patients

The spleen is capable of reducing red cell survival whenever red cell sequestration is enhanced. The susceptibility of antibody-coated red blood cells to splenic destruction has been previously mentioned. This is related to the special ability of the splenic reticuloendothelial cells to recognize the Fc fragment of IgG and the C3 component of complement. Red blood cells coated with either IgG or C3 attach to these surface receptors on the reticuloendothelial cells and are then phagocytized and destroyed.

The spleen also sequesters cells that contain precipitates of denatured hemoglobin, are misshappen, or are abnormally rigid. In some cases, the red blood cell can survive this filtering process. However, when it emerges from the spleen, it is often misshapen (poikilocytic or teardrop-shaped) and smaller. Removal of the spleen can slow the rate of red blood cell destruction by permitting these abnormal cells to remain in circulation. In the patient with an autoimmune hemolytic anemia, splenectomy may also reduce the level of antibody production.

Splenectomy has been used in the management of a number of congenital hemolytic anemias. The role of the spleen in hereditary spherocytosis is of particular importance because the increased destruction of red blood cells can be virtually eliminated by splenectomy. Therefore, hereditary spherocytic patients who are symptomatic (i.e., who complain of a lack of energy or easy fatigability that cannot be explained by another disease process) should be offered the option of splenectomy. Most of these patients experience a benefit. Unfortunately, success is not always guaranteed, and removal of the spleen is associated with some increase in the risk of serious infections such as pneumococcal or hemophilus septicemia. This complication is largely a matter of concern in childhood; adults are at much lower risk. All patients who are to be splenectomized should, however, be immunized with both polyvalent pneumococcus and hemophilus vaccines. They should also be counseled regarding the early treatment of any febrile illness.[105]

Hematologic improvement following splenectomy depends on the proportion of red blood cell destruction occurring in the spleen versus extrasplenic reticuloendothelial tissue. As a rule of thumb, patients with significant splenomegaly, secondary to work hypertrophy, will benefit from the procedure. When there is a question as to the role of the spleen, a direct assessment of red blood cell uptake can be determined by measuring the amount of radioactivity accumulating in the spleen after the transfusion of 51 chromium-tagged red blood cells. The performance and interpretation of this study are difficult at best and should be performed by an experienced nuclear-medicine facility. Regardless of the disease process, most patients show some uptake by extrasplenic reticuloendothelial tissue. Therefore, the decision to remove the spleen is usually based on the nature of the disease process, the patient's red blood cell transfusion requirement, and the size of the spleen. A rising transfusion requirement is often the clinical factor that tips the scale toward splenectomy.

Treatment of Iron Overload

Iron overload is a predictable complication of chronic red blood cell transfusion therapy. It also results from an increased level of iron absorption in patients with the genetic abnormality, idiopathic hemochromatosis, and in patients with erythropoietic disorders such as thalassemia and sideroblastic anemia, in whom ineffective erythropoiesis appears to stimulate excessive iron absorption. In each instance, excess iron is stored as ferritin and hemosiderin, but the location may vary. With multiple transfusions, the iron stores in reticuloendothelial cells increase dramatically. The parenchymal cells of the liver and other tissues are spared for a considerable period, until the gatekeeping role of the reticuloendothelial cell is overwhelmed and the serum iron increases. In idiopathic hemochromatosis, the reticuloendothelial cells are spared, and iron is deposited in a wide range of tissues. Organs at special risk for iron damage include the anterior pituitary, myocardium, pancreas, and liver. It is the parenchymal iron deposition that produces tissue damage, including anterior pituitary failure, heart failure, diabetes, and cirrhosis.

Thalassemia and ring sideroblastic anemia patients demonstrate both reticuloendothelial-cell and parenchymal-cell iron loading. The latter reflects these patients' high serum iron and full saturation of the iron-binding capacity. In addition, these patients have a high level of red cell destruction secondary to the ineffective erythropoiesis and reduced survival of circulating red blood cells. When the patient requires chronic red blood cell transfusion, the situation is made even worse. Patients in this category require aggressive treatment, for they can experience an even more rapid onset of tissue damage than that seen with idiopathic hemochromatosis.

Patients with idiopathic hemochromatosis having a functional erythroid marrow and a hemoglobin of 9 g/dL or more are best treated with weekly or every-other-week phlebotomy. As red blood cell iron is removed (1.1 mg/mL of red blood cells), a corresponding amount of iron is mobilized from storage areas. In idiopathic hemochromatosis, up to 500 mL of blood is removed each week until laboratory evidence of iron depletion appears. In a patient who presents with a serum ferritin level in excess of 3000 to 5000 μg/L, 6 months or more of treatment will be necessary before the serum ferritin level approaches the normal range. When the patient demonstrates an erythropoietic disorder, the anemia and/or the need for red blood cell transfusions will contraindicate phlebotomy therapy. In this situation, iron chelate therapy must be used. Desferrioxamine is a well-proven iron chelator, which must be administered parenterally to be effective. Ordinarily, this is done by subcutaneous pump infusion, 2 g over an 8-hour period during the night. Larger amounts of the drug, up to 8 g in a 24-hour period, can be administered by central venous catheter in patients who are severely ill and need rapid iron removal. The desferrioxamine is taken up by the hepatocyte, in which it binds to iron and is subsequently excreted into the bile. At lower doses, most of the iron chelate complex is reabsorbed and excreted in the urine. At high doses, it is excreted in both urine and stool. Each gram of chelate infused results in a loss of 25 mg of iron.

Treatment of Polycythemia

The first consideration in the treatment of a patient with an increased hemoglobin concentration is to determine whether this represents a true increase in red cell mass. If it does not, attempts to reduce the red cell mass are inappropriate. Indeed, in the elderly individual who presents with an increased hematocrit due to a decreased plasma volume, phlebotomy is hazardous. Attention should be given instead to the reestablishment of an adequate plasma volume.

Management of an individual who does have an increase in red cell mass varies according to the etiology of the disease process—whether it has occurred in response to hypoxia, is caused by a faulty regulation of EPO stimulation, or represents autonomous erythropoiesis. With hypoxia, a hematocrit increase to levels in excess of 60% is often associated with symptoms and signs related to increased blood viscosity. In this situation, phlebotomy to reduce the hematocrit to between 50% and 55% may be beneficial. The most desirable course of action, of course, is to reverse the underlying disturbance and thereby improve the patient's oxygenation. However, this is not always possible. Therefore, the physician must be guided by the severity of the hypoxia and the polycythemia, and most of all by the symptoms of the patient.

If there is excessive EPO stimulation, the specific cause must be determined. Particular attention is given to the possibility of a neoplasm, such as a hypernephroma, which can be surgically removed. If the cause cannot be corrected, phlebotomy is recommended to normalize the hemoglobin concentration. In younger patients with polycythemia vera, phlebotomy is the treatment of choice to normalize both the hemoglobin concentration and the total blood volume.[88] For older individuals, radioactive phosphorus or myelosuppressive chemotherapy may be indicated, particularly when there is a marked increase in platelets or granulocytes.

Whenever phlebotomy is undertaken in a polycythemic state, the initial goal is to control hypervolemia and decreased blood viscosity. As much as 1000 mL may be removed during the first week, followed by weekly phlebotomies of 500 mL. An early concern is the ability of the patient to withstand the transient hypovolemia associated with the phlebotomy. Depending on the patient's cardiovascular status, it can help to infuse 1 to 2 liters of saline immediately following the phlebotomy, to maintain the blood volume until the plasma volume can compensate.

Over the long term, phlebotomy will control the hemoglobin concentration by creating a state of iron deficiency. Usually, four to eight phlebotomies deplete iron stores. Therefore, after a few weeks or months, red cell regeneration depends solely on the level of iron absorption. Depending on the availability of iron in a normal diet, a 500-mL phlebotomy once every 3 to 4 months is usually sufficient to maintain the hemoglobin concentration in a normal range. However, tissue manifestations of iron deficiency may appear within months or years. Although this generally presents little problem to the patient, a short course of oral iron may be necessary. The patient must be warned not to take oral iron supplements at other times, because such supplements will result in a dramatic increase in the red cell mass.

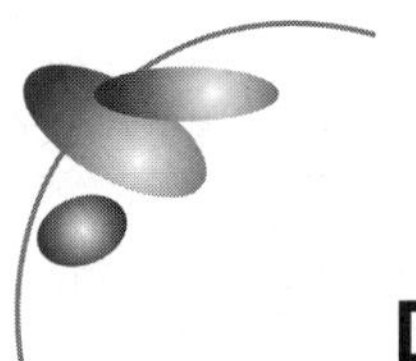

References

1. Fordham, EW and Ali, A: Radionuclide imaging of bone marrow. Semin Hematol 18:222, 1981.
2. Lichtman, MS: The ultrastructure of the hematopoietic environment of the marrow: A review. Exp Hematol 9:391, 1981.
3. Le Charpentier, Y and Prenant, M: Isolement de l'ilot erythroblastique. Nouv Rev Fr Hematol 15:119, 1975.
4. Ogawa, M, Porter, PN, and Nakahata, T: Renewal and commitment to differentiation of hemopoietic stem cells: An interpretive review. Blood 61:823, 1983.
5. Kapoff, CT and Jandl, JH: Blood: Atlas and Source Book of Hematology. Little, Brown & Co., Boston, 1981.
6. Papayannopoulou, T and Finch, CA: Radioiron measurements of red cell maturation. Blood Cells 1:535, 1975.
7. Goldwasser, E: Erythropoietin and the differentiation of red blood cells. Fed Proc 34:2285, 1975.
8. Finch, CA: Erythropoiesis, erythropoietin, and iron. Blood 60:1241, 1982.
9. Chanarin, I, et al: Cobalamin-folate interrelationships: A critical review. Blood 66:479, 1985.
10. Finch, CA, et al: Erythrokinetics in pernicious anemia. Blood 9:807, 1956.
11. Morgan, EH: Transferrin biochemistry, physiology and clinical significance. Mol Aspects Med 4:1, 1981.
12. Gidati, AS and Lever, RD: Enzymatic formation and cellular regulation of heme synthesis. Semin Hematol 14:145, 1977.
13. Bessis, M: Living blood cells and their ultrastructure. (Translated by RI Weed.) Springer-Verlag, New York, 1973.
14. Chasis, JA and Shohet, SB: Red cell biochemical anatomy and membrane properties. Ann Rev Physiol 49:237, 1987.
15. Palek, J (ed): Blood cell cytoskeleton. I. Red cell membrane skeleton. Semin Hematol 20:139, 1983.
16. Marcus, D (ed): Blood group immunochemistry and genetics. Semin Hematol 18:1, 1981.
17. Ballas, SK, et al: Red cell membrane and cation deficiency in Rh null syndrome. Blood 63:1046, 1984.
18. Jaffe, ER and Heller, P: Methemoglobinemia in man. In Brown, EG and Moore, CV (eds): Progress in Hematology, Vol IV. Grune & Stratton, New York, 1964, p. 48.
19. Cohen, RJ, et al: Methemoglobinemia provoked by malarial chemoprophylaxia in Vietnam. N Engl J Med 279:1127, 1968.

20. Oski, FA and Gottlieb, AJ: The interrelationships between red blood cell metabolites, hemoglobin, and the oxygen equilibrium curve. In Brown, EG and Moore, CV (eds): Progress in Hematology, Vol. VII. Grune & Stratton, New York, 1971, p 33.
21. Perutz, MD: The Croonian Lecture, 1968. The haemoglobin molecule. Proc R Soc Lond 173:113, 1969.
22. Eichner, ER: Splenic function: Normal, too much and too little. Am J Med 66:311, 1979.
23. Bunn, HF: Erythrocyte destruction and hemoglobin catabolism. Semin Hematol 9:317, 1972.
24. Bunn, HF and Jandl, JH: The renal handling of hemoglobin. J Exp Med 129:925, 1969.
25. Krantz, SB: Erythropoietin. Blood 77:419, 1991.
26. Bauer, C and Kurtz. A: Oxygen sensing in the kidney and its relation to erythropoietin formation. Ann Rev Physiol 51:845, 1989.
27. Finch, CA and Lenfant, C: Oxygen transport in man. N Engl J Med 286:407, 1972.
28. Wade, OL and Bishop, JD: Cardiac Output and Regional Blood Flow. Blackwell Scientific Publications, Oxford, 1962.
29. Giblett, ER, et al: Erythrokinetics. Quantitative measurements of red cell production and destruction in normal subjects and patients with anemia. Blood 11:291, 1956.
30. Beguin, Y, et al: Ferrokinetic measurement of erythropoiesis. Acta Haematol 79:121, 1988.
31. Intraquintornchai, T, et al: In vivo transferrin-iron receptor relationships in erythron of rats. Am J Physiol 255:326, 1988.
32. Hillman, RS: Characteristics of marrow production and reticulocyte maturation in normal man in response to anemia. J Clin Invest 48:443, 1969.
33. Horne, MK, III, et al: "Early-Peak" carbon monoxide production in certain erythropoietic disorders. Blood 45:365, 1975.
34. Berlin, N and Berk, PD: Quantitative aspects of bilirubin metabolism for hematologists: A review. Blood 57:983, 1981.
35. International Committee for Standardization in Haematology: Recommended method for radioisotope red-cell survival studies. Br J Haematol 45:659, 1980.
36. Hillman, RS and Henderson, PA: Control of marrow production by relative iron supply. J Clin Invest 48:454, 1969.
37. Finch, CA, et al: Ferrokinetics in man. Medicine 49:17, 1970.
38. Dawson, AA, Ogston, D, and Fullerton, HW: Evaluation of diagnostic significance of certain symptoms and physical signs in anaemic patients. Br Med J 436, 23 August 1969.
39. Duke, M and Abelmann, WH: The hemodynamic response to chronic anemia. Circulation 39:503, 1969.
40. Card, RT and Brain, MD: The "anemia" of childhood. N Engl J Med 288:388, 1973.
41. Garn, SM, et al: Income matched black-white hemoglobin differences after correction for low transferrin saturations. Am J Clin Nutr 34:1645, 1981.
42. Bessman, JD, Gilmer, PR, and Gardner, FH: Improved classification of anemias by MCV and RDW. Am J Clin Pathol 80:322, 1983.
43. Westerman, MP: Bone marrow needle biopsy: An evaluation and critique. Semin Hematol 18:293, 1981.
44. Stevens, AR, Coleman, DH, and Finch, CA: Iron metabolism: Clinical evaluation of iron stores. Ann Intern Med 38:199, 1953.

45. Bainton, DF and Finch, CA: The diagnosis of iron deficiency anemia. Am J Med 37:62, 1964.
46. Gilmer, PR and Koepke, JA: The reticulocyte. Am J Clin Pathol 66:262, 1976.
47. Labardini, J, et al: Marrow radioiron kinetics. Haematologica 7:301, 1973.
48. Perrotta, AL and Finch, CA: The polychromatophilic erythrocyte. Am J Clin Pathol 57:471, 1972.
49. Lipschitz, DA, Cook, JD, and Finch, CA: A clinical evaluation of serum ferritin as an index of iron stores. N Engl J Med 290:1213, 1974.
50. Cook, JR: Clinical evaluation of iron deficiency. Semin Hematol 19:6, 1982.
51. Finch, CA and Huebers, H: Perspectives in iron metabolism. N Engl J Med 306:1520, 1982.
52. Guyatt, GH, et al: Laboratory diagnosis of iron-deficiency anemia. J Gen Intern Med 7:145, 1992.
53. Means, RT and Krantz, SB: Progress in understanding the pathogenesis of the anemia of chronic disease. Blood 80:1639, 1992.
54. Faguin, WC, Schneider, JJ, and Goldberg, MA: Effect of inflammatory cytokines on hypoxia-induced erythropoietin production. Blood 79:1987, 1992.
55. Adamson, JW, Eschbach, J, and Finch, CA: The kidney and erythropoiesis. Am J Med 44:725, 1968.
56. Tudhope, GR: Endocrine diseases. Clin Haematol 1:475, 1972.
57. Viteri, FE, et al: Hematological changes in protein calorie malnutrition. Vitam Horm 26:573, 1968.
58. Adamson, JN and Finch, CA: Hemoglobin function, oxygen affinity and erythropoietin. Ann Rev Physiol 37:351, 1975.
59. Dessypris, EN: Aplastic anemia and pure red cell aplasia. Curr Opin Hematol 1:157, 1994.
60. Williams, DM, Lynch, RE, and Cartwright, GE: Drug-induced aplastic anemia. Semin Hematol 10:195, 1973.
61. Weatherall, DJ: The thalassemias. In Beutler, E, et al (eds): Williams Hematology, ed 5. McGraw-Hill, New York, 1995.
62. Kushner, JP, et al: Idiopathic sideroblastic anemia: Clinical and laboratory investigation of 17 patients and review of the literature. Medicine 50:139, 1971.
63. Solomon, LR, Hillman, RS, and Finch, CA: Serum ferritin in refractory anemia. Acta Hematol 66:1, 1981.
64. Stabler, SP, Allen, RH, Savage, DG, and Lindenbaum, J: Clinical spectrum and diagnosis of cobalamin deficiency. Blood 76:871, 1990.
65. Eichner, ER, Pierce, I, and Hillman, RS: Folate balance in dietary induced megaloblastic anemia. N Engl J Med 284:933, 1971.
66. Steinberg, SE, Campbell, CL, and Hillman, RS: Kinetics of the folate enterohepatic cycle. J Clin Invest 64:83, 1979.
67. Hillman, RS and Steinberg, SE: The effects of alcohol on folate metabolism. Ann Rev Med 33:345, 1982.
68. Eichner, ER and Hillman, RS: The evolution of anemia in alcoholic patients. Am J Med 50:218, 1971.
69. Lindenbaum, J, Savage, DG, Stabler, SP, and Allen, RH: Diagnosis of cobalamin deficiency II. Relative sensitivities of serum cobalamin, methylmalonic acid and homocysteine concentrations. Am J Hematol 34:99, 1990.
70. Cooper, BA and Rosenblatt, DS: Inherited defects of vitamin B_{12} metabolism. Ann Rev Nutr 7:291, 1987.
71. Marchand, A, Galen, RS, and Van Lente, F: The predictive value of serum haptoglobin. JAMA 243:1909, 1980.
72. Beutler, E: G6PD deficiency. Blood 84:3613, 1994.

73. Nydegger, UE, Kazatchkine, MD, and Miescher, PA: Immunopathologic and clinical features of hemolytic anemia due to cold agglutinins. Semin Hematol 28:66, 1991.
74. Rosse, WF and Adams, JP: The variability of hemolysis in the cold agglutinin syndrome. Blood 56:409, 1980.
75. Palek, J and Sahr, KE: Mutations of the red blood cell membrane proteins: From clinical evaluation to detection of the underlying genetic defect. Blood 80:308, 1992.
76. Rosse, WF: Paroxysmal nocturnal hemoglobinuria. In Clinical Immunohematology: Basic Concepts and Clinical Applications, ed 1. Blackwell Scientific Publications, Oxford, England, 1990.
77. York, EL, et al: Effects of secondary polycythemia on cerebral blood flow in chronic obstructive pulmonary disease. Am Rev Respir Dis 121:813, 1980.
78. Wedzicha, JA, et al: Erythrapheresis in patients with polycythaemia secondary to hypoxic lung disease. Br Med J 286:511, 1983.
79. Harrison, BDW and Stokes, TC: Secondary polycyhthemia: Its causes, effects and treatment. Br J Dis Chest 76:313, 1982.
80. Watts, EJ and Lewis, SM: Spurious polycythemia. A study of 35 patients. Scand J Haematol 31:241, 1983.
81. Pearson, TC, Glass, UH, and Wetherly-Mein, G: Interpretation of measured red cell mass in the diagnosis of polycythemia. Scand J Haematol 21:153, 1978.
82. Wynne, JW, Block, AJ, and Boysen, PG, Jr: Oxygen desaturation in sleep: Sleep apnea and COPD. Hosp Pract, October 1980, p 77.
83. Annotation: Polycythemia and high-affinity haemoglobins. Br J Haematol 36:153, 1977.
84. Basu, TK and Stein, RM: Erythropoiesis associated with chronic renal disease. Arch Intern Med 133:442, 1974.
85. Thorling, EB: Paraneoplastic erythrocytosis and inappropriate erythropoietin production. A review. Scand J Haematol Supp. 17, 1972.
86. Adamson, JW: The erythropoietin/hematocrit relationship in normal and polycythemic man: Implications of marrow regulation. Blood 32:597, 1968.
87. Adamson,JW, et al: Recessive familial erythrocytosis: Aspects of marrow regulation in two families. Blood 41:641, 1973.
88. Golde, DW, et al: Polycythemia: Mechanisms and management. Ann Intern Med 95:71, 1981.
89. Eridani, S, et al: Erythroid colony formation in primary proliferative polycythemia, idiopathic erythrocytosis and secondary polycythaemia: Sensitivity to erythropoietic stimulating factors. Clin Lab Haematol 5:121, 1983.
90. Shine, KI, et al: Aspects of the management of shock. Ann Intern Med 93:723, 1980.
91. Maier, RV and Carrico, CJ: Developments in the resuscitation of critically ill surgical patients. Adv Surg 19:271, 1986.
92. Blumberg, N, et al: Immune response to chronic red blood cell transfusion. Vox Sang 44:212, 1983.
93. Eschbach, JW and Adamson, JW: Recombinant human erythropoietin: Implications for nephrology. Am J Kidney Dis 11:203, 1988.
94. Nissenson, AR, Nimer, SD, and Wolcott, DL: Recombinant human erythropoietin and renal anemia: Molecular biology, clinical efficacy, and nervous system effects. Ann Intern Med 114:402, 1991.
95. Fischl, M, et al: Recombinant human erythropoietin therapy for AIDS patients treated with AZT: A double-blind, placebo-controlled clinical study. N Engl J Med 322:1488, 1990.

96. Goodnough, LT, et al: Increased preoperative collection of autologous blood with recombinant human erythropoietin therapy. N Engl J Med 321:1163, 1989.
97. Lederle, FA: Oral cobalamin for pernicious anemia: Medicine's best kept secret? JAMA 265:94, 1991.
98. Hillman, RS, Adamson, J, and Burka, E: Characteristics of B_{12} correction of the abnormal erythropoiesis of pernicious anemia. Blood 31:419, 1968.
99. Crump, M, et al: Treatment of adults with severe aplastic anemia: Primary therapy with antithymocyte globulin (ATG) and rescue of ATG failures with bone marrow transplantation. Am J Med 92:596, 1992.
100. Gluckman, E, et al: Results of immunosuppression in 170 cases of severe aplastic anaemia. Br J Haematol 51:541, 1982.
101. Storb, R, et al: One-hundred-ten patients with aplastic anemia (AA) treated by bone marrow transplantation in Seattle. Transplant Proc 10:135, 1978.
102. Collins, PW and Newland, AC: Treatment modalities of autoimmune blood disorders. Semin Hematol 29:64, 1992.
103. Goldberg, MA, et al: Treatment of sickle cell anemia with hydroxyurea and erythropoietin. N Engl J Med 323:366, 1990.
104. Charache, S, et al: Effect of hydroxyurea on the frequency of painful crises in sickle cell anemia. N Engl J Med 332:1317, 1995.
105. Schwartz, PE, et al: Postsplenectomy sepsis and mortality in adults. JAMA 248:2279, 1982.

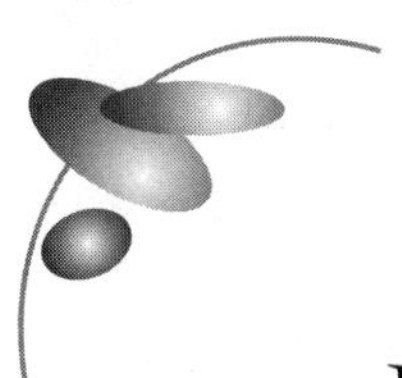

Index

An "f" following a number indicates a figure. A "t" indicates a table. "CP" indicates a color plate.